POSTHARVEST MANAGEMENT OF HORTICULTURAL CROPS

POSTHARVEST MANAGEMENT OF HORTICULTURAL CROPS

Editor

Jeltsyn Rocha

scitus
academics

Postharvest Management of Horticultural Crops
Edited by **Jeltsyn Rocha**

ISBN: 978-1-68117-704-5
Library of Congress Control Number: 2016958744

SCITUS Academics LLC,
616, Corporate Way, Suite 2, 4766,
Valley Cottage, NY 10989

www.scitusacademics.com

Notice

Printed in United States of America on Acid Free Paper

PREFACE

Post-harvest priorities worldwide have substantially evolved in the last four decades, from being purely technical in their outlook to being much more responsive to customer demand. Consumer trends that have contributed to this shift include rising incomes in urban areas, changing nutritional habits, reduced demand times and preparing meals for safety, comfort and quality. Other aspects, such as globalization, urbanization and the need to achieve efficiencies and reduce costs, have significantly contributed to this change in priorities and continue to shape and restructure the fresh produce industry.

Post-harvest management consists of the various practices and technologies undertaken by farmers, farmers ' cooperatives or groups and/or agribusiness enterprises, from yield to plate, to manage crop production immediately after harvest, as well as its final destination, such as storage, sorting, cleaning, transport, packaging and processing. The main objectives of post-harvest handling are to keep the product cool, to stay away from loss of moisture and to slow down unwanted chemical changes and to stay away from physical damage such as bruising or cutting, to delay spoilage.

Combining several pre- and post-harvest strategies to change these physiological activities, leading to an increase in shelf life, the book also discusses awesome technologies that have a positive impact on the quality of the products, including senescenal changes and the consumer's decision to purchase the product on the market. The book includes chapters covering detailed discussions on post-harvest management strategies for horticultural commodities with contributions from industry experts with experience in both developed and developing regions. Post-harvest management obviously determines the quality and safety of food, the competitiveness of the market and the profits earned by producers. This will include a much closer and more systematic link between operations and stakeholders, the modernization of marketing infrastructure and technologies, the capacity building of individual stakeholders and the strengthening of policy/institutional framework for better marketing. All sectors, like the public and private sectors, need to work together to mitigate these constraints.

TABLE OF CONTENTS

1

PLANT EXTRACTS IN POST-HARVEST DISEASE MANAGEMENT OF FRUITS AND VEGETABLES-A REVIEW

Anjum Malik A[1], Naseer Ahmed[2*], Babita[3], Harmeet Chauhan[1]and Prerna Gupta[1]

[1]Division of Post-Harvest Technology, Sher-e-Kashmir university of Agricultural Science and Technology, Jammu, India

[2]Choudhary Charan Singh Hisar Agriculture University, Hissar, Haryana, India

[3]Department of Fruit Science, Dr. Y S Parmar University of Horticulture and Forestry, Nauni-Solan, HP, India

ABSTRACT

Purpose of research: This review focus of the utilization of plant extracts in the post harvest management of fruits and vegetables and this also emphasize the need for future research. **Findings:** Plant products are an important source of agrochemicals used for the control various post harvest loses which include diseases as well as insect pests. The widely studied plants in this context are the neem tree (Azadirachta indica), chinaberry (Melia azadrach) and marigold (Tagetes spp). They are being used to manufacture natural or bio insecticides, which are environmental friendly and do not have any toxic effects on plants and soil. More ever they possess fungicidal and insecticidal properties. Various extracts such as neem leaf extract, castor oil and neem oil on citrus fruits and reported that, among these extracts neem was best in retaining most of biochemical characteristics such as TSS (16.01°B), acidity (0.38%), pectin (0.98%) and ascorbic acid content (20.56 mg/100 ml juice). **Directions for future research:** Each of these categories requires further research. The industries would benefit from this article on the utilization of plant extract post harvest management of fruits and vegetables. There are several areas which require further exploration. One of these is the development of technologies for the extraction of plant extract and its awareness about the safety of its utilization without any effect on human health.

Keywords: Neem; Chinaberry; Marigold; Post harvest management

INTRODUCTION

Although satisfactory control of the disease by using various chemicals have been documented in the literature [1-3] during recent years, global concern for protection of the environment has led researchers to investigate the use of natural flora as one of the sources for crop protection [4].

Antimicrobial activity of water and ethanolic extracts from five medicinal plants against bacteria isolated from disease fairy shrimps. The antimicrobial efficacy of *Andrographis paniculata, Garcinia mangostana, Psidium guajava, Punica granatum* and *Terninalia catappa* was evaluated against four gram negative bacteria including*Aeromonas sobria, Klebsiella pneumonia, Enterobacter sp. and Chryseobacterium sp.* using agar well diffusion method at three concentration levels (100, 10 and 1 mg/ml). Among the tested plants, *Punica granatum* extract showed the highest level of antimicrobial activity against all four bacterial strains at 10 mg/ml. The ethanolic extract of Punica granatum was very active against *Aeromonas sobria* at a concentration of 3.125 mg/ml

A considerable amount of work has been carried out all over the world on the effect of plant extracts on the quality and shelf life in various fruit and vegetable crops. Fungal infections are one of the major causes of post harvest rots of fresh fruits and vegetables whether in transit or storage. They cause significant economic losses in the commercialization phase,and are rendered unfit for human consumption. Neem (*Azadirachta indica* A. Juss) and chinaberry (*Melia azadrach*) are two trees belonging to the mahogany family (Meliaceae). Neem is an evergreen tree [5], whereas chinaberry is a deciduous tree native to South Asia Australia. Fishwick [6] and Eeswara [7] studied the reputed value for its antifungal properties. Siddiqui et al. [8] evaluated the efficacy of leaf extract of neem (*Azadirachta indica*) and Chinaberry (*Melia azadrach*) against two tomato pathogenic fungi *Alternaria solani* and*Fusarium oxysporium*, the causal agents of early blight and wilt disease of tomato plants respectively were studied. Leaf extracts of different concentration (5%, 10%, 15% and 20%) of aqueous, ethanol and ethyl acetate of neem and chinaberry were added to growth media prior to inoculation. Inhibition values recorded characteristic variation between neem and chinaberry. In case of neem inhibition percentages were 17.88%, 23.66%, 52.77% and 70.50%against *Alternaria solani*. While for *F. oxysporum* were 14.77%, 23.88%, 31.22% and 100% for all four concentrations used. Whereas for chinaberry inhibition values for *Alt. solani* is3.11%, 5.22%, 5.53%, and 5.77% and 5.44%, 6.11%, 6.35% and 6.55% for *F. oxysporum* [9]. The efficacy of the two plant extracts against the tomato fruit rot fungi was tested in vitro. The results showed that, the extracts significantly ($P < 0.05$) inhibited the mycelial growth of the fungal pathogens at the different concentrations tested and the rate of inhibition differed from one extract to the other. Ilondu et al. [10] reported that some plants contain phenolic substances and essential oils, which are inhibitory to micro-organisms. The presence of these compounds in these extracts has been reported to be responsible for their antifungal properties [11]. These antifungal properties control various pests including fungi while the extract of ginger rhizomes is specially valued for their effectiveness against fungi [11].

LITERATURE REVIEW

Azadirachtin is considered as the most active principal substance in neem which has growth regulating, fungicidal and insecticidal properties. Biopesticides are a good alternative to the synthetic pesticide. Both leaves and fruit of neem plant are known to have bitter taste having fungicidal, insecticidal and nematicidal properties. Azadirachtin, chemically a tetranorterpenoid component of neem acts on the mitotic cells and blocks the microtubule polymerization. Certain activities of genes and proteins are also altered by azadirachtin. As a botanical insecticide azadirachtin is effective against many biological processes. It may cause a reduction in feeding habit, suspend the molting process, larvae and pupae death and also cause sterility in the emerging adults, this all depend on the given dose. Apart from azadirachtin, many other components of neem also have insecticidal properties. Among them are saponins that are found to have antimicrobial activity that inhibits moulds and protecting plant from attack of insects [12]. Many biologically active compounds have been isolatedfrom neem, *Azadirachta indica* A. Juss, including triterpenoids, azadirachtin [13] and melantriol [14]. This reduced the feeding of insects. Azadirachtin is a mixture of seven isomeric compounds as Azadirachtin-A to Azadirachtin-G of which Azadirachtin-E is the most effective insect growth regulator [15]. Azadirachtin possess insecticidal, ovicidal, antifeedant and growth inhibiting effects against many insect pests [16-18]. Azadirachtin has minimal impact on nontarget organisms and is compatible with other eco-friendly bio control agents [19]. Effects of a botanical pesticide, azadirachtin, were evaluated on non-target aquatic macroinvertebrates found in SW Montana. A neem product formulated for commercial use and its petroleum-based, inert carrier alone were used to test the hypothesis that a commercial, U.S. Environmental Protection Agency-approved formulation of the pesticide, azadirachtin-A, derived from extracts of neem tree seeds,*Azadirachta indica* A. Juss. (Family: Meliaceae), is toxic to nontarget aquatic macroinvertebrates in Montana river systems. LC50 s (24 h) for 5 insect species, Drunella grandis (McDunnough), Drunella doddsi (Needham), Skwala parallela (Frison), Brachycentrus occidentalis Banks, and Brachycentrus americanus (Banks), and 1 isopod species, Caecidotea inter- media (Forbes) varied from 1.8 ppm to

9.2 ppm of the full formulation. The full formulation consisted of: azadirachtin (3% wt:wt); other neem-extracted material (32% wt/wt); petroleum solvents (65% wt:wt). LT50 s for 4 ppm of this formulation was 9.4 to 748 hours for these peld-collected macro- invertebrates. Without azadirachtin, but with the petroleum-based carrier, LT50 s was 13.6 h to 46.7 h. These bioassays conducted late in winter and early spring of 1994 and 1995 and late summer and early fall of 1994 indicated mortality from the neem formulation was similar to herbicides 2,4-D and picloram, but mortality could be accounted for by action of petroleum solvents in the inert materials. This suggests the registration process of new pesticides should also include documented potential toxicity of inert components in the formulation to non-target aquatic insects [20]. Marigold belongs to the genus Tagetes (Asteraceae) with 56 species and includes the popular ornamental budding plants known as marigolds. Deshmukh et al. [21] have reported the beneficial effect of Tagetes erecta flower on various horticultural and agricultural crops, which are attributed primarily to the presence of an organic compound 'camacin' exhibiting antifungal and insecticidal effect. Chauhan et al. [22] treated apple cv. Starking delicious with plant leaves/flower based formulation of neem leaf extract (10%, 20%), spearmint (Mentha spicata) leaf extract (10%, 20%), marigold (Tagetes erecta) flower extract (10%, 20%) and semper fresh (control 1.5%). Freshly harvested fruits were treated with above treatments and were kept under refrigerated storage (1 ± 1°C) for analysis at a month interval up to 180 days. Among treatments, 20% spearmint leaf extract proved to be highly effective in reducing fruit spoilage followed by 20% neem leaf extract. Fruit treated with 20% drake leaf extract proved to be most effective treatment in reducing weight loss and also 20% drake and 20% neem leaf extract were capable to retain maximum total soluble solids.

Pandey et al. [23] applied leaf extracts of *Mentha* and *Melia* to guava fruits and observed that these treatments were effective in retaining physical quality characteristics of fruits as they reduced moisture loss and storage rot symptoms. Rameshwar et al. [24] also observed that a combination of neem leaf extract at 20% + UV radiation exposure for 10 minutes + rice starch 6% proved to be the most appropriate treatment in minimizing the reduction of juice contents and fruit rotting and fluctuation in physiological loss in weight was also minimum as compared to control.

Hasabnis and D' souza [25] reported beneficial effects of natural plant products for the control of storage rots in Alphonso mango. They further reported that these treatments also checked the incidence of disease development in banana. Neem is the most promising and widely used against a number of insect pests due to its versatile mode of action. It has a broad spectrum action on insects and is very useful. However, the problems of quality, shelf life, presence of aflatoxins and phytotoxicity limit its efficacy and eco-safet. Rameshwar et al. [26] observed that a combination of neem leaf extract at 20% + UV radiation exposure for 10 minutes + rice starch 6% proved to be the most appropriate treatment in minimizing the reduction of juice contents and fruit rotting and fluctuation in physiological loss in weight was also minimum as compared to control. Wijewardane and Guleria [19] studied about the concentration (1.5-2%) of neem oil as a surface coating along with pre-cooling at 10°C was the most effective in providing a better physico-**chemical** umbrella, significantly lowering the fruit spoilage. The neem oil surface coating along with shrink-wrapped tray packing at 2°C (85% to 90% RH) resulted in the longest storage life up to 150 days. Among the rates tested, the highest retention of fruit firmness, and pectin content, anthocyanin pigments, and polygalacturonase enzyme activity in fruits was recorded by 20%marigold extract, surface coated with shrink-wrapped tray packing along with pre-cooling. Arya [26] reported that higher concentration (75%) leaf extract of neem was effective in reducing spores (82.3%) of two fruit rot pathogens, *Phomopsis psidii* and *Phomopsis viticola*. From**Table 1** the pre and post inoculation rot could not be checked completely by any plant extract but, the disease severity was reduced with all the plant extracts in comparison to control [27].

Chauhan and Joshi [22] reported the efficacy of phytoextracts on the storage quality of mango cv. Ratna and found them significantly better in retaining total soluble solids and sugar contents and in reducing reduction in the possible incidence of anthracnose pathogen in comparison to untreated fruits where lower soluble solid and sugar content and higher incidence of anthracnose pathogen was reported. The increase in TSS and sugar content may be due to the hydrolysis of insoluble polysaccharides into simple sugars. Such changes are expected to be slower and more gradual when the metabolism of the commodity is slowed down by the application of various coating treatments, pre-cooling and under low temperature storage.

Table 1: Table showing the effect of per inoculation and post-inoculation treatments on severity of Aspergillus rot in Aonla fruits.

	Fruit Rot Severity (%)			
	Pre-Inoculation		Post-Inoculation	
Treatment	**Days after inoculation (DAI) 3**	**(DAI) 6**	**(DAI) 3**	**(DAI) 6**
Aloe Vera	2.75	4.25	3.25	4.75
Jatropha	0.00	2.75	0.00	3.00
Neem seed kernel	3.25	5.75	3.50	6.00
Garlic	0.00	2.35	0.00	2.50

Tiwari [28] Rao and Singh [29] have reported beneficial effect of various botanical extracts such as neem leaf extract, *Melia* leaf extract, onion extract, garlic extract, mahua extract and eucalyptus extract on the post-harvest quality of fruits and vegetables. They are also reported to act as antifeedent and antirepellent agents against various storage pathogens. Rameshwar et al. [24] observed that a combination of neem leaf extract at 20% + UV radiation exposure for 10 minutes + rice starch 6% proved to be the most appropriate treatment in minimizing the reduction of juice contents and fruit rotting and fluctuation in physiological loss in weight was also minimum as compared to control.

Deshmukh et al. [21] have reported beneficial effects of *Melia (Melia Azadirachta)* and bavistin in controlling the storage fungi and insect-pests. They have also advocated commercial use of *Melia* extract as a substitute for fungicides and insecticides. *Melia azedarach L.* extracts were studied in comparison with selected biorational insecticides against the citrus leaf miner Phyllocnistis citrella Stainton under field conditions. Citrus limon (L.) Burm. F. trees were exposed to: *Melia* extracts of green and mature fruits, Neem oil (30% a.i.), abamectin (1.8% a.i.) and control. Two sprays of each treatment (except for *Melia* mature fruit extract) were executed at 10-day intervals. The live number of the 1st and later (2nd & 3rd) larval instars per leaf were recorded at initial sampling date and at 10-day intervals after each spray application. Results indicated that there were significant differences in the number of live larval instars among treatments. *Melia* extracts and the two biorationals, neem oil and abamectin, decreased the larvae population significantly to lower numbers than that of the control at 10 days after each spray application. However, the decrease caused by neem oil and abamectin was significantly higher than that of *Melia* extracts. Thus, these extracts might be considered as potential alternative with other biorational control methods in management of the leaf miner. Further research including bioassays is needed to determine the factors responsible for reducing larvae population and whether these *Melia* extracts can be utilized in future citrus IPM programs as a tool for citrus leaf miner management [30].

Lehmann and Ibenthal [31], Bottenburg and Singh [32] reported beneficial effects of treatments with neem leaf and mentha leaf extract on the marketing quality characteristics of cow pea and sugar beet and concluded that these treatments were effective in retaining the physicochemical characteristics of these crops. The TSS, acidity, phenols and ascorbic acid showed less decrease as compare to the control. In control the degradation of quality is more as compare to that of treated with neem leaf and menthe leaf extract. Sarvamangla [33] studied the effect of leaf extracts from *Azadirachta indica, Calotropis gigantean, Calthranthus sp. Eucalyptus sp., Parthenium chysterophorus*and *Pongamia pinnata* on fruit quality of mulberry and reported that among these extracts *Azadirachta indica* was the most effective in retaining higher sugar content, maintaining fruit size and shape, and resistance to fungal pathogens. *Azadirachta indica* control the fungal pathogen more effectively and retaining the maximum sugar content as compare to *Calotropis gigantean, Calthranthus sp. Eucalyptus sp., Parthenium chysterophorus* and*Pongamia pinnata* (**Table 2**).

Srivastava and Lal [34] reported that leaf extract of *Azadirachta indica* was effective in the control of fruit rot of pear and pomegranate. Sindhan et al. [35] reported the beneficial effects of neem, eucalyptus, tulsi, datura, bougainvillea and ginger on the bio-chemical and physical quality characteristics of citrus and mango fruits, as these extracts significantly reduced **moisture** loss and retained higher soluble solid content over the uncoated fruits. Rameshwar et al. [24] observed that a combination of neem leaf extract at 20% + UV radiation exposure for 10 minutes + rice starch 6% proved to be the most appropriate treatment in minimizing the reduction of juice contents and fruit rotting and fluctuation in physiological loss in weight was also minimum as compared to control (Table 2). Bhardwaj and Sen [36] observed that PLW, rotting, loss in juice content and reduction in diameter of fruits of sweet orange cv. Nagpur Santra under zero energy cool chamber + 20% neem leaf extract were significantly less, whereas high under ambient condition + 10% neem leaf extract after 42 days of storage. Borthakar et al. [37] conducted experiments on the effects of neem leaf (5%), melia leaf (5%) garlic (5%) and neem oil (2%) on physicochemical characteristics of Baramasi lemon fruit and concluded that among various treatments neem leaf extract and melia leaf extract was better in retaining most of the physico-chemical characteristics of fruits.

Chauhan et al. [38] treated apple cv. Starking Delicious with plant leaves/flower based formulation of neem leaf extract (10%, 20%), spearmint (*Mentha spicata*) leaf extract (10% and 20%), marigold (Tagetes erecta) flower extract (10%, 20%) and semper fresh (control 1.5%). Freshly harvested fruits were treated with above treatments and were kept under refrigerated storage (1 ± 1°C) for analysis at a month interval up to 180 days. Among treatments, 20% spearmint leaf extract proved to be highly effective in reducing fruit spoilage followed by 20% neem leaf extract. Fruit treated with 20% drake leaf extract proved to be most effective treatment in reducing weight loss and also 20% drake and 20% neem leaf extract were capable to retain maximum total soluble solids. Similar findings have been reported by Ali et al. [39], Abid and Maqbool [40] on tomato fruits.

Bhardwaj and Sen [37] studied the effect of different concentrations of neem leaf extracts on the storage quality of mandarin (*Citrus reticulata*) cv. Nagpur Santra and reported that among various treatments used neem leaf extract (20%) was significantly better in retaining higher ascorbic acid content (27.17 mg/100 ml. of juice) as compared to control fruits where it was only after storage. Chaudhary [41] reported that aqueous extracts from different plants can be a viable alternative to chemicals in controlling post-harvest pathogens as they are environmentally safe strategies for controlling the post- harvest storage pathogens in different crops. Singh et al. [42] also reported the effect of various extracts such as neem leaf extract, castor **oil** and neem oil on citrus fruits and reported that, among these extracts neem was best in retaining most of biochemical characteristics such as TSS (16.01° B), acidity (0.38%), pectin (0.98%) and ascorbic acid content (20.56 mg/100 ml juice) as compared to control fruits in which the values for these parameters were 12.03° B, 0.23%, 0.55% and 15.68 mg/100 ml juice, respectively after 12 days of storage.

Table 2: Table showing the effect of per inoculation and post-inoculation treatments on severity of Aspergillus rot in Aonla fruits.

Name	Plant species	Family name	Nature of the extract	Property
Neem	*Azadirachta indica L.*	Meliaceae	Leaf powder	Anti-fungal
Hot chili	*Capsicum frutescence L.*	Solanaceae	Fruit powder	Anti-fungal
Lemon grass	*Cymbopogon nardus L.*	Poaceae	Leaf powder	Anti-fungal
Turmeric	*Curcuma longa L.*	Zingiberaceae	Tuber powder	Anti-fungal
Ginger	*Zingiber officinale L.*	Zingiberaceae	Tuber powder	Anti-fungal

Singh et al. [43] observed that the quality of untreated mango cv. Langra fruit started deteriorating after about two weeks of storage and the fruits became unacceptable after five weeks. However, the fruits were reported to be acceptable even after five weeks of storage when they were given post-harvest dip of GA3 (500 ppm) and 20% neem leaf extract and retained high pectin content (0.98%) over the control (0.55%) fruits. Singh et al. [42] tested the efficacy of aqueous leaf extracts of 20 plant species and 3 essential oils against management of *Penicillium expansum* rot of apple cv. Red Delicious. Among various treatments leaf extract of *Azadirachta indica* significantly reduced P. expansum rot when used as post-infection dip treatment. Chauhan et al. [38] treated apple cv. Starking Delicious with plant leaves/ flower based formulation of neem leaf extract (10%, 20%), spearmint (Mentha spicata) leaf extract (10%, 20%), marigold (*Tagetes erecta*) flower extract (10%, 20%) and semper fresh (control 1.5%). Freshly harvested fruits were treated with above treatments and were kept under refrigerated storage (1 ± 1°C) for analysis at a month interval up to 180 days. Among treatments, 20% spearmint leaf extract proved to be highly effective in reducing fruit spoilage followed by 20% neem leaf extract. Fruit treated with 20% drake leaf extract proved to be most effective treatment in reducing weight loss and also 20% drake and 20% neem leaf extract were capable to retain maximum total soluble solids. Hassanein et al. [44] studied the effectiveness of leaf extracts of neem and china berry against two tomato pathogenic fungi *Alternaria solani* and *Fusasuim oxysporum*, the causal agent of early blight and wilt disease of tomato plants respectively and reported promising results. Jakhar et al. [45] reported that the physiological loss in weight, colour, texture, shriveling and rotting of Chakaiya cultivar of aonla fruit was significantly affected by packaging materials, neem leaf extracts and diphenyl fumigation and their combination throughout the storage period up to 21 days.

Patel et al. [46] reported that the extract from *Allium sativum* was significantly better followed by extract of *Jatropha curcas*, Aloe barbadensis and Azadirachta indica in controlling Aspergillus niger of aonla fruit rot. Okigbo et al. [47] used ethanolic and **water** extractions of *Azadirachta indica* leaves and *Aframomum melegueta* seeds as antifungal agents against pathogenic fungi *Aspergillus niger, Botryodiplodia theobromae, Fusarium solani* and *Penicillium oxalicum* of cassava and reported that *A. indica* was more active on the organisms than *A. melegueta* in preventing post-harvest deterioration of cassava. Bhardwaj et al. [48] reported that Azadiracta indica and Chromoleana odorata were able to suppress rot-causing fungi of yam. Therefore, serve as a good natural plant fungicide (protectant) against yam tubers in storage.

CONCLUSION

Plant extracts are being used to manufacture natural or bio insecticides, which are environmental friendly and do not have any toxic effects on plants and soil. More ever they possess fungicidal and insecticidal properties. So these can be used as compare to synthetic/ chemical products. Various extracts such as neem leaf extract castor oil and neem oil on citrus fruits and reported that, among these extracts neem was best in retaining most of biochemical characteristics such as TSS, acidity, pectin and ascorbic acid content. Moreover the extract of neem, marigold and other plant extracts increase the shelf life of the fruits and vegetables by reducing the fungal and bacterial spoilage during **storage**. Neem extracts was best in retaining most of biochemical characteristics such as TSS, acidity, pectin and ascorbic acid content. Plant extracts are important in order to reduce the environmental pollution by synthetic chemicals. In day to day life the world is facing the threat of global warming and pollution, in order to contribute in keeping the environment clean to some extent and reduce the residual effect of synthetic insecticide and pesticides.

REFERENCES

1. Choulwar AB, Datar VV (1988) Cost linked spray scheduling for management of tomato early blight. Indian Phytopathol 41: 603-606.

2. Maheshwari SK, Gupta PC,Gandhi SK (1991) Evaluation of different fungi toxicants against early blight of tomato. Agriculture Science Digest 11: 201-202

3. Abdul Mallek AY, Hemida SK, Bagy MK(1995) Studies on fungi associated with tomato fruits and effectiveness of some commercial fungicides against three pathogens. Mycopathologia 130: 109-116.

4. Wijewardane RMNA, Guleria SPS(2009) Combined effect of Pre-cooling, application of natural extracts and packaging on the storage quality of Apple (*Malus domestica*) cv. Royal Delicious. Tropical Agricultural Research21: 10-20.

5. Schmutterer H(1995) The Neem Tree. Publisher VCH verlagsgesells chaft weinheim (Federal Republic of Germany) and VCH Publishers Inc., New York, USA.

6. Fishwick RW (1989) Neem (*Azadirachta indica* A. Juss) Plantations in the Sudan Zone of Nigeria Rep.Prepared for the Chief Conservator of Forests, Northern Nigeria.

7. Eeswara JP, EJ Allan, AA Powell (1998) The influence of storage of seed maturity, moisture content and storage temperature on the survival of neem (*Azadirachta indica*) seed in storage. Seed Sci Technol 26: 299- 308.

8. Siddiqui BS, Afshan F, Gulzar T, Hanif M (2004) Tetracyclic triterpenoids from the leaves of *Azadirachta indica*. Phytochemistry 65: 2363-2367.

9. Hassanein NM, Abou Zeid MA, Youseef KA, Mahmoud DA (2008) Efficacy of neem (*Azadirachta indica*) and C hinaberry (*Melia azedrach*) against early blight and wilt of tomato. Aus JBasic Appl Sci 2: 763-772.

10. Ilondu EM, Ejechi BO, Souzey JA (2001) Microbial stability of jam prepared from velvet tamarind and preserved bycombined processes. Nig J Microbiol 5: 93-96.

11. Ahmed S, Stoll G (1996) Biopesticides. In: Haverkort JB, Hiemstra W (eds.)Biotechnology: Building on Farmers' Knowledge. Macmillan Education Ltd,Bunders, London.

12. Bajwa AA, Ahmad A (2012) Potential applications of Neem based products as biopesticides. The Health 3: 116-120.

13. Butterworth JH, Morgan ED (1971) Investigation of the locust feeding inhibition of the seeds of the neem tree,*Azadirachta indica*. J Insect Physiol17: 969-977.

14. Lavie D, Jain MK, Shpan-Gabrielith SR (1967) A locust phago-repellent from two media species. ChemComm 13: 910-911.

15. Verkerk RHJ, Wright DJ (1993) Biological activity of neem seed kernel extracts and synthetic azadirachtin against larvae of *Plutella xylostella* L. Pestic Sci37: 83-91.

16. Akou-Edi D (1984) Effects of neem seed powder and oil on Tribolium confusum and Sitophilus zeamais. In: Schmutterer H, Ascher KRS (eds.) Natural pesticides from the neem tree (*Azadirachta indica* A. Juss) and other tropical plants. Rauischholzhausen, Federal Republic of Germany, pp: 445-452.

17. Schmutterer H (1990) Properties and potential of natural pesticides from the neem tree, *Azadirachta indica*. Annu Rev Entomol 35: 271-297.

18. Nawrot J, HarmathaJ (1994) Natural products as antifeedants against stored products insects. PostharvNews Inf 5: 17-21.

19. Wijewardane RMNA, Guleria SPS (2009) Combined Effects of Pre-cooling, Application of Natural Extracts and Packaging on the Storage Quality of Apple (*Malus domestica*) cv. Royal Delicious 21: 10-20.

20. Florence V, Dunkel, David CR (2008) Effect of an Azadirachtin Formulation on Six Nontarget Aquatic Macroinvertebrates; Environ. Entomol 27: 667-674.

21. Deshmukh PB, Chavam SR, Rehapukar DM (1992) A study of insecticidal activity of twenty indigenous plants. Pesticide 16: 7-10.

22. Chauhan HI, Joshi HN (1990) Evaluation of phyto extracts for control of mango anthracnose. Proceedings of Symposium of Botanical Pesticides.

23. Pandey RS, Bhargava SN, Shukla DN, Dwivedi KK (1983) Control of pestololia fruit rot of guava by leaf extracts of two medicinal plants. Revists Mexic Fito 2:15-16.

24. Rameshwar, Dayal, Jeewa, Ram, Shafeal, et al. (1987) Role of Post-harvest treatment with plant extracts, ultra violet radiation and rice starch on keeping quality of Sweet Orange (*Citrus sinesis osbeck*) cv. mosambi. J Eco Physiol.

25. Arya A (1988) Control of phomosis fruit rot by leaf extract of certain medicinal plants. In: Kaushik P (eds.). Indigenous Medicinal Plants. Symposium. Today and Tomorrows Printers and Publishers, New Delhi, pp: 41-46.

26. Hasabnis SN, D'Souza TF (1987) Use of natural products in the control of the storage rot in alphanso mango fruits. J Maharashtra Agri Univer 12:105-106.

27. Patel DS, Nath K, Patel RL, Patel SI (2008) Management of fruit rot of Aonla caused by *Aspergillus niger*. 38: 658-651.

28. Tiwari DN (1991) Monograph on neem (*Azadirachta indica*). International book distribution New Delhi, India.

29. Rao PJ, Singh S (1999) Effect of Artemisia vulgaris on the development of various pests during storage. Indian J Entomol 61: 305-314.

30. Lehmann W. and Ibenthal WA. Fungicidal effects of neem extracts against seedling blight of sugar bet. Symposium on useful Tropical Plants, Hamburg, Germany.

31. Maher MM, Efat MA, Fakhr Hammad, Mohamad TF (2013) Effect of Melia zedarach (*Sapindales*: Meliaceae) fruit extracts on Citrus Leafminer Phyllocnistis citrella(*Lepidoptera*: Gracillariidae) 2: 144

32. Bottenburg BH, Singh BD (1996) Effect of neem leaf extract applied using the broom method on cow pea pests and yield. Int J Pest Manag42:207-209.

33. Sarvamangla HS (1993) Evaluation of plant extract for the control of fungal disease of blackberry. Indian Phytopath 46: 398-401.

34. Srivastava AK, Lal P (1997) Studies on bio fungicidal properties of leaf extracts of some plant. Indian Phytopath50: 408-411.

35. Sindhan GS, Hooda, Indira, Prashar RD (1999) Effect of some plant extracts on the vegetative growth and storage rot causing fungi. J MycolPlant Pathol29: 110-111.

36. Bhardwaj RL, Sen NL (2003) Physico-chemical changes of stored mandarin orange (*Citrus reticulata Blanco*) cv. 'Nagpur Santra' as affected by neem leaf extract and zero energy cool chamber. South Indian Horti 50: 500-504.

37. Borthakar PK, Ranjit K, Kumar R (2002) Effect of plant extract on size and rind thickness of Baramasi lemon fruit (Citrus lemon) during storage. J Agri SciSoc North-East India 15:98-103.

38. Chauhan, Sandeep K, Thakur KS, Bhanot, Amandeep, et al. (2008) Effect of pre and postharvest treatment of neem (*Azadirachta indica*) based formulations, plant leaves and their extracts on the storage quality of apple (*Malus domestica*). Indian J Agri Sci78: 966-970.

39. Abid M, Maqbool MA (1991) Effect of bare-root dip treatment in oil cakes of neem leaf extract on the root-knot development and growth of tomato and eggplant. Pak J Nematol 9: 13-16.

40. Ali TE, Nasir MA, Shakir AS (1992) In-vitro evaluation of certain neem products as mould inhibitors against post-harvest rotting fungi of tomato. Pak J Phytopathol4:58-61.

41. Chaudhary M (2003) Studies on the post-harvest diseases of tomato and their management. Dr. Y.S. Parmar University of Horticulture and Forestry, Nauni-Solan (HP).

42. Singh, Dinesh, Thakur RK, Singh D (2003) Effect of pre harvest sprays of fungicides and calcium nitrate on post-harvest rot of kinnow in low temperature storage. Plant Dis Res18: 9-11.

43. Singh JN, Acharya P, Singh BB (2000) Effect of GA3 and plant extracts on storage behavior of mango (Mangifera indica) cv Langra. Har J Horti Sci29: 3-4.

44. Hassanein MM, Abou Zeid MA, Youssef KA, Mahmoud DA (2008) Efficacy of leaf extracts of neem (*Azadirachta indica*) and chinaberry (*Melia azedrach*) against early blight and wilt disease of tomato. Aus J BasicApplSci 2: 763-772.

45. Patel DS, Nath K, Patel RL and Patel SI. Management of fruit rot of aonla caused by Aspergillus niger. JMycology Plant Pathol38: 658-660.

46. Jakhar RP, Dheeraj Singh (2008) Post-harvest effect of packaging material, neem leaf extract and fumigation on color, texture, shriveling and rotting of aonla (*Emblica officinalis*). Environ Eco 26: 1652-1656.

47. Bhardwaj RL, Dhashora LK, Mukherjee S (2010) Effect of plant extract and benzyladenine on post-harvest shelf life of orange (*Citrus reticulata Blanco*). J Adv Dev Res1: 32-37.

2

ADVANCES IN FRUIT AROMA VOLATILE RESEARCH

Muna Ahmed Mohamed El Hadi, Feng-Jie Zhang, Fei-Fei Wu, Chun-Hua Zhou * and Jun Tao

College of Horticulture and Plant Protection, Yangzhou University, Jiangsu Key Laboratory of Crop Genetics and Physiology, Yangzhou 225009, China

ABSTRACT

Fruits produce a range of volatile compounds that make up their characteristic aromas and contribute to their flavor. Fruit volatile compounds are mainly comprised of esters, alcohols, aldehydes, ketones, lactones, terpenoids and apocarotenoids. Many factors affect volatile composition, including the genetic makeup, degree of maturity, environmental conditions, postharvest handling and storage. There are several pathways involved in volatile biosynthesis starting from lipids, amino acids, terpenoids and carotenoids. Once the basic skeletons are produced via these pathways, the diversity of volatiles is achieved via additional modification reactions such as acylation, methylation, oxidation/reduction and cyclic ring closure. In this paper, we review the composition of fruit aroma, the characteristic aroma compounds of several representative fruits, the factors affecting aroma volatile, and the biosynthetic pathways of volatile aroma compounds. We anticipate that this review would provide some critical information for profound research on fruit aroma components and their manipulation during development and storage.

Keywords: fruit; aroma volatiles; pre & postharvest; biosynthetic pathways

1. INTRODUCTION

Fruit quality includes both its preharvest development, such as changes in colour, flavor, and texture as fruits develop, grow, and ripen, as well as its maintenance following harvest as the perishable tissues senesce [1]. Flavor consists both of the perception in mouth (sweetness, acidity or bitterness) and on the odor, produced by several volatile compounds. All plants are able to emit volatile organic compounds (VOCs) and the content and composition of these molecules show both genotypic variation and phenotypic plasticity [2]. As aroma is one of the most appreciated fruit characteristics, volatile flavor compounds are likely to play a key role in determining the perception and acceptability of products by consumers. Identification of key volatile flavor metabolites that carry the unique character of the natural fruit is essential, as it provides the principal sensory identity and characteristic flavor of the fruit [3].

Aroma is a complex mixture of a large number of volatile compounds, whose composition is specific to species and often to the variety of fruit [4,5]. Although different fruits often share many aromatic characteristics, each fruit has a distinctive aroma that depends upon the combination of volatiles, the concentration and the perception threshold of individual volatile compounds [6]. The most important aroma compounds include amino acid-derived compounds, lipid-derived compounds, phenolic derivatives, and mono- and sesquiterpenes [5]. Although fruit aroma is generally a complex mixture of a wide range of compounds, volatile esters often represent the major contribution in apple (*Malaus domestica* Borkh.) and peach (*Prunus persica* L.) [7,8].

As an important trait of fruit quality, aroma has gained increasing attention in recent years. With the fast development of science and technology especially the application of the GC-MS and other analytical apparatus, progress in aroma research has been made in several fields [9]. In this review, we summarized the composition of fruit aroma and the characteristic aroma compounds of several representative fruits at first, then the factors affecting aroma volatile were discussed, and finally the biosynthetic pathways of volatile aroma compounds were summarized.

2. AROMA VOLATILE COMPOSITION AND THEIR BIOLOGICAL CHARACTERISTIC OF MAJOR FRUITS

Most fruits produce significant numbers of volatile compounds as indicators of fruit ripening. Many of these volatile compounds are produced in trace amounts, which are below the thresholds of most analytical instruments, but can be detected by human olfaction [10]. Volatiles can be classified as primary or secondary compounds, indicating whether they were present in intact fruit tissue or produced as a result of tissue disruption [11]. It should be pointed out that analysis of volatiles from either intact or disrupted fruit tissues will influence the aroma profiles and final aroma interpretation. The volatile profiles of fruit are complex and vary depending on the cultivar, ripeness, pre-and post-harvest environmental conditions, fruit sample (either intact fruit, slices, or homogenized samples), and analytical methods utilized [12,13]. Aroma compounds are often only released upon cell disruption when previously compartmentalized enzymes and substrates interact [14]. Some aroma compounds are bound to sugars as glycosides or glucosinolates. Glycosides of aroma compounds in fruit are mainly *O*-*β*-D-glucosides and *O*-diglycosides, but triglycosides have also been identified [15]. The proportion of glycosidically bound volatiles is usually greater than that of free volatiles, making them an important potential source of flavor compounds. The odorous aglycones may be released from the sugar moiety during maturation, processing and storage, or by the action of enzymes, acids or heat [16].

2.1. Classification of Volatile Compounds in Fruit Flavor

Various types of fresh fruits produce distinct volatile profiles. Flavor volatiles are derived from an array of compounds including phytonutrients such as fatty acids, amino acids, carotenoids, phenols and terpenoids [10]. Fruit volatile compounds are mainly comprised of diverse classes of chemicals, including esters, alcohols, aldehydes, ketones, lactones, and terpenoids. However, some sulfur compounds, such as *S*-methyl thiobutanoate, 3-(methylthio) propanal, 2-(methylthio) ethyl acetate, 3-(methylthio) ethyl propanoate, and 3-(methylthio) propyl acetate, also contribute to the flavor of fruit such as melons (*Cucumis melo* L.) [9]. Bound volatiles are recognized as a potential source of aroma compounds in fruits such as kiwifruit (*Actinidiadeliciosa*) [17]. Although an overwhelming number of chemical compounds have been detected as volatile compounds in fresh fruit, only a fraction of these compounds have been identified as impact components of fruit flavor based on their quantitative abundance and olfactory thresholds [18]. Many C_{10} monoterpenes and C_{15} sesquiterpenes compose the most abundant group of compounds present in the aroma profile. In some cases, these are also the key compounds determining the characteristic aroma. For example, the terpenoids *S*-linalool, limonene, valencene and β-pinene are key aroma compounds of strawberry (*Fragaria x ananassa*), koubo (*Cereus peruvianus* L.) and citrus (*Citrus* sp.) [19,20,21]. Volatile terpenoid compounds, potentially derived from carotenoids, are important components of flavor and aroma in many fruits. Of particular interest are a group of terpenoid flavor volatile compounds generally present at relatively low levels but possessing strong effects on the overall human appreciation. Among these are β-ionone, geranylacetone (6,10-dimethyl-5,9-undecadien-2-one), pseudoionone (6,10-dimethyl-3,5,9-undecatrien-2-one), β-cyclocitral, geranial, theaspirone, α-damascenone and β-damascenone. Their structures reveal an isoprenoid-based origin, and they were long assumed to be the products of the oxidative cleavage of carotenoids [22].

2.2. Volatile Compounds and Their Biological Characteristic of Major Fruits

More than 300 volatile molecules have been reported in fresh apples [23]. The total number, identity and concentration of volatile compounds emitted by ripening apple fruit are cultivar specific [24]. The contribution of each compound to the specific aroma profile of each cultivar depends on the activity and substrate specificity of the relevant enzymes in the biosynthetic pathway, the substrate availability, the odour threshold above which the compound can be detected by smell, and the presence of other compounds [25]. Esters are the most abundant volatile compounds emitted by apple and, together with α-farnesene, have been proposed for cultivar classification [26]. Ethyl 2-methyl butanoate, 2-methyl butyl acetate, and hexyl acetate contribute mostly to the characteristic aroma of "Fuji" apples, while ethyl butanoate and ethyl 2-methyl butanoate are the active odor compounds in "Elstar" apples, and ethyl butanoate, acetaldehyde, 2-methyl butanol, and ethyl methyl propanoate in "Cox Orange" [27]. For the "Pink Lady" cultivar, hexyl acetate, hexyl 2-methyl butanoate, hexyl hexanoate, hexyl butanoate, 2-methylbutyl acetate and butyl acetate were prominent within the blend of volatiles produced by fruit throughout maturation [28]. Among fruit tissues, it has been shown that epidermal tissue produces a greater amount of volatiles than internal tissues [29]. This higher capacity for aroma production by the peel has been attributed to either the abundance of fatty acid substrates or the higher metaboltic activity [30,31].

More than 300 volatile compounds have been identified in pear fruit (*Pyrus pyrifolia* Nak.) [32], methyl and hexyl esters of decadienoate are the character-imparting compounds of the European pear [32,33]. Other volatile esters, for example, hexyl acetate, 2-methylpropyl acetate, butyl acetate, butyl butanoate, pentyl acetate, and ethyl hexanoate possess strong pear-like aromas [32]. Ethyl octanoate and ethyl-(*E*)-2-octenoate contribute to sweet or fruity odors in pears, while a high concentration of 2,4-decadienoates in fruit flesh is accepted by consumers [34]. In addition, hexanal, 2-methylpropyl acetate, ethyl acetate, hexyl acetate, 3-methylbutyl-2-methyl butanoate, ethyl butanoate, and butanol are identified as impact volatiles in "Conference" pears [35].

Although the melon aroma has been extensively investigated and more than 240 volatile compounds have been identified in different varieties [36], the literature is lacking in quantitative data. Numerous compounds of different degrees of volatility, especially those containing a C_9 straight chain, are the major determinants of melon fruit quality perceived by consumers. These compounds are strongly dependent on the variety and physiological behavior of the fruit; in fact, fresh climacteric melons such as cantaloupe have greater aroma intensity and a shorter shelf life than less climacteric melons such as honeydew [37]. In climacteric aromatic melon varieties, volatile esters are prominent, together with sulphur-containing aroma compounds, sesquiterpenes, norisoprenes, short-chain alcohols, and aldehydes [38]. Volatile esters, mainly acetate derivatives such as ethyl 2-methyl propyl acetate and 2-methyl butyl acetate are dominant with 37% of the total volatile profile [39]. In addition, lower amounts of lactones, sulfur compounds [such as (methylthio) acetate, 2-(methylthio) ethyl acetate and 3-(methylthio) propyl acetate], short-chain alcohols and aldehydes compose the complex mixture of volatile compounds [40,41,42]. Non-aromatic varieties often have much lower levels of total volatiles, and lack the volatile esters [43]. Volatiles derived from amino acids are major contributors to melon aroma [44,45]. Both aromatic and non-aromatic varieties possess amino acid derived volatiles. In the aromatic varieties these volatiles are mostly esterified and their levels are usually higher than in the non-aromatic varieties. In the non-aromatic varieties they occur as aldehydes and alcohols.

Strawberry has one of the most complex fruit aromas, made up of approximately 350 volatile compounds [5,46]. The furanones 2,5-dimethyl-4-hydroxy-3(2*H*)-furanone (furaneol) and its methyl derivative 2,5-dimethyl-4-methoxy-3(2*H*)-furanone (mesifurane) are considered the dominating aroma compounds [47]. They contribute to the typical caramel-like, sweet, floral and fruity aroma. Esters, which are the most important group of strawberry aroma compounds, cover 90% of the total number of volatiles in ripe strawberry fruit [47]. Among the major esters are methyl and ethyl butanoate, butyl acetate, methyl and ethyl hexanoate. Other key aroma compounds are linalool, γ-decalactone and 2,3-butanedione though some of these key compounds tend to be cultivar-specific [47,48]. Finally, aldehydes and alcohols such as hexanal,*trans*-2-hexenal and *cis*-3-hexen-1-ol are important for the green, unripe notes in strawberry aroma. Their concentrations are also cultivar and ripeness dependent [47].

Although more than 250 volatile components have been identified in banana [49], the banana fruity top notes are from volatile esters, such as isoamyl acetate and isobutyl acetate [50]. The concentrations of acetates and butanoates increased during ripening of banana fruit [49]. In addition, isoamyl alcohol, isoamyl acetate, butyl acetate, and elemicine were detected by olfactometric analyses as characteristics of banana odor [51]. The main volatile compounds found in "Cavendish" banana were (*E*)-2-hexenal and acetoin, in "Plantain": (*E*)-2-hexenal and hexanal, and in "Frayssinette": 2,3-butanediol and two diastereomeric solerols. The most abundant aglycones were 3-methyl butanol, 3-methyl butanoic acid, solerol (two diastereomers) and acetovanillone. This compound, rarely identified in fruits, was detected for the first time in conjugated volatile compounds of fruits. The abundance of these two diastereomer in the extracts of "Frayssinette" seemed to be characteristic of this variety of banana [52].

Citrus volatiles have been extensively examined over the past several decades. Esters are important as they are responsible for the flavor characteristic [27].While the major esters are ethyl esters of C_3 to C_4organic acids, linalool is by far the most important alcohol. However, ketones, carvone, diacetyl, and acetoin are off-flavors. Volatile compounds in citrus fruits accumulate in oil glands of flavedo and in the oil bodies of the juice sacs. Over 100 volatile compounds have been identified for the first time in the juice of four varieties of citrus (Powell Navel orange (*Citrus sinensis* L.), Clemenules mandarine (*C. reticulate* Blanco.), Fortune mandarine (*C. reticulate* Blanco.) and Chandler pummelo (*C. maxima* Merr.)). The differences in the volatile profile in citrus juice are mainly quantitative, and only a few compounds are variety-specific. In

Chandler the most characteristic volatiles were principally aliphatic aldehydes, sesquiterpenes such as nootkatone and monoterpenes such as 2-carene. Powell Navel orange showed the highest levels of esters such as nonyl acetate and of monoterpenes such as 3-carene. Clemenules showed the highest levels of ketones 3-pentanone and β-ionone and Fortune showed the highest levels of some acetate esters such as ethyl and propyl acetate, this latter almost Fortune-exclusive [53]. Moreover, a total of 58 volatile components were identified and quantified in Dortyol yerli orange juice, terpenes quantitatively and qualitatively represent the main group of the volatile fraction. D,L-limonene was the major terpene component, followed by valencene. After terpenes, terpenols were clearly the dominant constituents. Linalool, terpinen-4-ol and α-terpineol were the most abundant among the terpenol compounds. In terms of aroma contribution to orange juice, 12 compounds were prominent based on the odour activity values (OAVs). The highest OAV values were recorded for ethyl butanoate, nootkatone, linalool and *D,L*-limonene [54]. On the other hand, limonene was the most abundant monoterpene hydrocarbon in both the peel and juice of three major Japanese sour citrus cultivars, Yuzu (*C. junos* Sieb. ex Tanaka), Sudachi (*C sudachi* Hort. ex Shirai) and Kabosu (*C. sphaerocarpa* Hort. ex Tanaka), followed by γ-terpinene and β-phellandrene in Yuzu and Sudachi and by myrcene in Kabosu. Mintsulfide was newly identified in juice extracts from Yuzu and Kabosu. Among the oxygenated components, linalool was the most abundant in both the peel and juice of Yuzu and the peel of Sudachi, while both the peel and juice of Kabosu revealed the presence of high quantity of saturated aliphatic aldehydes. Wine lactone and rose oxide were identified in all the extracts, which have not been previously reported to occur in these citrus cultivars. With regard to linalool, the *R*-enantiomer was predominant in Yuzu and the *S*-enantiomer in Sudachi and Kabosu [55]. In other citrus species, such as *C. natsudaidai* other terpenoids are less abundant, but they exert a profound effect on aroma. These include γ-terpinene, β-phellandrene, mycerene, and α-pinene [20,56].

Mango (*Mangifera indica* L.) possesses a very attractive flavor characteristic. More than 270 aroma volatile compounds in different mango varieties have been identified in free form [57]. However, application of distillation extraction in combination with active odor value (aroma threshold) technologies exhibits that monoterpenes such as *α*-pinene, myrecene, *α*-phelladrene, *σ*-3-carene, *p*-cymene, limone and terpinolene, esters including ethyl-2-methyl propanaote, ethyl butanoate, as well as (*E,Z*)-2,6-nonadienal, (*E*)-2-nonenal, methyl benzoate, (*E*)-*β*-ionone, decanal, and 2,5-dimethyl-4-methoxy-3(2*H*)-furanone are the most important compounds contributing to mango flavor [58]. Generally, terpene hydrocarbons are the major class of compounds in New World mangoes, with contents ranging from 16–90%. 3-Carene is the major compound in most New World mango cultivars, with limonene, *β*-ocimene, myrcene and *α*-terpinolene having some importance in some cultivars. Sesquiterpene hydrocarbons may also be present in amounts as high as 10% in some cultivars. There is a large variation in the quality and quantity of alcohols, ketones, and esters in mangoes, especially those of the Old World varieties. Those compounds, together with esters, are responsible for much of the characteristic aroma of Old World mangoes [59]. Terpene hydrocarbons were the major volatiles of nine varieties of Colombian mango, the dominant terpenes being δ-3-carene ("Haden", "Irwin", "Manila" and "Tommy Atkins"), α-pinene ("Hilacha" and "Vallenato"), α-phellandrene ("Van Dyke") and terpinolene ("Yulima"), while no dominant terpene was found in the fruit of "Springfield" [60]. Aroma volatile compounds in the mango fruit have also been reported to be present in the glycosidically-bound form [61].

Peach fruit volatiles have been extensively studied, and more than 100 compounds were identified [62,63,64], among them, C_6 aldehydes and alcohols offer the green-note aroma, while lactones and esters are responsible for fruity aromas. Esters such as hexyl acetate and (*Z*)-3-hexenyl acetate are considered as key odorants influencing the flavor quality of peach fruit [65]. Changes in aroma-related volatiles have been reported during peach fruit development and postharvest ripening [66]. Aldehydes tend to decline, while esters increase in the fruit, and postharvest treatments, such as low temperature and controlled atmosphere, have been used to investigate postharvest changes in peach aroma quality [8,67].

To date, more than 200 different volatile compounds have been described in apricots (*Prunus armeniaca* L.) [68]. Six major volatile compounds were identified (hexanal, (*E*)-2-hexenal, linalool, 1-hexanol, ethyl octanoate, and hexyl acetate) in apricot fruit cv. "Modesto", all of which were previously reported to be major contributors to apricot aroma [69,70]. The most abundant volatile compounds in terms of concentration were aldehydes, mainly hexanal and (*E*)-2-hexenal, and their concentration decreased during ripening with significant differences. Terpenic compounds (i.e. linalool) and alcohols (i.e., 1-hexanol) were produced at a lower concentration than aldehydes, and decreased during ripening with a similar pattern to that observed for aldehydes [69].

Grape (*Vitis vinifera* L.) volatiles include a great number of compounds, among which monoterpenes, C_{13}norisoprenoids, alcohols, esters and carbonyls are found [71]. Grape may be divided into aromatic and nonaromatic varieties. Free terpenols, for example, linalool and geraniol, have been identified as major aroma compounds in both red and white grapes [72]. Octanoic acid and alcohols, particularly 2-phenylethanol, are recognized after crushing [72]. In addition, esters and aldehydes were also reported in "Aleatico" grapes [73]. The most abundant free compounds detected in muscat grape were linalool, geraniol, citronellol, nerol, 3,7-dimethyl-1,5-octadien-3,7-diol (diendiol I) and 3,7-dimethyl-1,7-octadien-3,6-diol (diendiol II). In the glycosidically-bound fraction the major compounds were geraniol, linalool, citral, nerol, citronellol, α-terpineol, diendiol I, diendiol II, *trans*-furan linalool oxide (linaloloxide I), *cis*-furan linalool oxide (linaloloxide II), benzyl alcohol and 2-phenylethanol. Other monoterpenes potentially contributing to Muscat aroma were rose oxide, citral, geraniol, nerol and citronellol [74]. (*E*)-2-hexenal was the most abundant volatile compound in Riesling and Cabernet Sauvignon grapes, and it showed a significant increase in concentration after veraison. Benzene derivatives discriminated ripe Cabernet Sauvignon grapes, whereas monoterpenes and sesquiterpenes discriminated both cultivars pre-veraison with a broader range of terpenes observed in the Cabernet Sauvignon samples compared with the Riesling samples. At veraison, terpene production in both varieties was low, but Riesling grapes produced some terpenes (geraniol and *α*-muurolene) post-veraison. Generally, esters and aldehydes were the major class of compounds from Riesling grapes, while Cabernet Sauvignon showed a greater tendency to form alcohols [75].

Aroma compounds in raspberry (*Rubus idaeus* x *ursinus*) have been studied extensively. At least 200 volatile compounds have been identified in this fruit. Many compounds including raspberry ketone, *α*-ionone, *β*-ionone, linalool, (*Z*)-3-hexenol, geraniol, nerol, *α*-terpineol, Furaneol, hexanal, *β*-ocimene, 1-octanol, *β*-pinene, *β*-damascenone, ethyl 2-methylpropanoate, (*E*)-2-hexenal, heptanal, and benzaldehyde have been identified to contribute raspberry aroma. Among them, *α*-ionone, *β*-ionone, geraniol, nerol, linalool, and raspberry ketone could be particularly important to red raspberry aroma [76]. Twenty-nine volatile compounds were quantified in "Chilliwack", "Tulameen", "Willamette", "Yellow Meeker", and "Meeker" raspberries, data showed that volatile concentrations varied for different cultivars. Large variations for *α*-ionone, *β*-ionone, geraniol, linalool, and (*Z*)-3-hexenol were observed in different raspberry cultivars. In addition, the volatile compositions in "Meeker" raspberry grown at different locations also varied [77]. Although blackberry (*Rubus laciniata*) has been widely planted, the study of blackberry flavor is still very limited. The early studies focused on the volatile constituents of blackberry, and very diverse compounds have been identified. 2-Heptanol, *p*-cymen-8-ol, 2-heptanone, 1-hexanol, *α*-terpineol, pulegone, 1-octanol, isoborneol, myrtenol, 4-terpineol, carvone, elemicine, and nonanal have all been identified as the major volatiles [78]. In order to understand the aroma differences, the volatile compositions of "Marion" and "Black Diamond" was analyzed using stir bar sorptive extraction-gas chromatography–mass spectrometry (SBSE-GC–MS) and solid phase extraction (SPE)-microvial insert thermal desorption-GC–MS for two growing seasons. Although seasonal variations were present, the overall volatile profile in "Marion" and "Black Diamond" were very similar, but the concentrations of some aroma compounds varied greatly. Odour-activity value (OAV) indicated that furaneol, linalool, *β*-ionone, and hexanal could be most important in "Marion", while in "Black Diamond", the most important compounds were linalool, *β*-ionone, furaneol, and 2-heptanol. The major difference between the cultivars for aroma compounds was that "Marion" had higher OAV of furaneol than "Black Diamond", while "Black Diamond" had higher OAV of linalool than "Marion". The chemical analysis results matched with the descriptive sensory evaluation that "Marion" had more berry, fruity, strawberry aroma while "Black Diamond" had more floral aroma [79].

A total of 42 volatiles were identified in four southern highbush blueberry (*Vaccinium sp.*)cultivars ("Primadonna", "Jewel", "Snowchaser", and "FL02-40") , twelve of these volatiles are reported for the first time in highbush blueberries, with 10 being positively identified: (*Z*)-3-hexenal, (*E*,*E*)-2,4-hexadienal, (*E*,*Z*)-2,6-nonadienal, (*E*,*E*)-2,4-nonedienal, methyl 2-methylbutanoate, butyl acetate, 2-methylbutyl acetate, and geranyl acetate. "Primadonna" was characterized by a large amount of esters and C_6 aldehydes. In contrast, fewer than four esters were found in "FL02-40" and "Snowchaser", respectively, but they produced more terpenoids than "Primadonna" and "Jewel" [80].

More than 280 volatile compounds have been found in pineapple fruit (*Ananas comosus* L. Merr.) [81]. Esters and hydrocarbons were the major constituents. Octenoic acid, methyl ester, hexanoic acid, octanoic acid and ethyl ester were the specific aromatic components of pineapple fruits. The relative content of volatiles showed significant variations during the fruit developmental period [82]. A total of 11 and 28 volatile compounds were identified in the Tainong No. 4 and No. 6 pineapples, respectively. According to the OAVs, four compounds were defined as the characteristic aroma compounds for the Tainong No. 4 pineapple, including furaneol, 3-(methylthio) propanoic acid methyl ester, 3-(methylthio) propanoic acid ethyl ester and δ-octalactone. The OAVs of five compounds including ethyl-2-methylbutyrate, methyl-2-methylbutyrate, 3-(methylthio) propanoic acid ethyl ester, ethyl hexanoate and decanal were considered to be the characteristic aroma compounds for the Tainong No. 6 pineapple [83]. The characteristic aroma compounds in the plup of cayenne pineapple were ethyl 2-methylbutanoate, ethyl hexanoate, 2,5-dimethyl-4-hydroxy-3(2*H*)-furanone (DMHF), decanal, ethyl 3-(methylthio) propionate, ethyl butanoate, and (*E*)-3-ethyl hexenoate;

while in core the main compounds were ethyl-2-methyl butanoate, ethyl hexanoate and DMHF. The highest odor units were found to correspond to ethyl 2-methyl butanoate, followed by ethyl hexanoate and DMHF [84].

More than 80 compounds related to fruit aroma have been identified in kiwifruit, with the major components being methyl and ethyl butanoate, (*Z*)- and (*E*)-2-hexenal, hexanal, (*Z*)- and (*E*)-3-hexenol, and methyl benzoate [17]. The volatile profile of eating-ripe "Hort16A" is dominated by ethyl esters and 3-(methylsulfanyl) ethyl propionate has been suggested to contribute to the tropical aroma of non-stored kiwifruit [85,86]. "Hort16A" has been less extensively studied, but it has been noted that an important difference between the aroma profiles of "Hort16A" and "Hayward" is the presence of diverse sulphur compounds in the former [17].

3. FACTORS INFLUENCING VOLATILE COMPOSITION

Due to the complex nature of the volatile profiles, volatile composition is continuously changing in fresh fruit. Many factors affect volatile composition including the genetic makeup of the fruit, its maturity, environmental conditions during production, postharvest handling, and storage. To date we have a limited understanding of how these factors interact to determine the actual volatile composition and resulting flavor of the fruit.

3.1. Genetics

Evaluation of volatiles at the germplasm level is useful for future breeding efforts, aimed at improvement of fruit quality, via effects on fruit aroma. The composition and concentration of grape volatiles largely varied with genetic background. In grape cultivars belonging to seven genotypic groups, C_6 compounds were the dominant volatiles in *Vitis. amurensis* grape, neutral cultivars *V. vinifera* grapes and hybrid grapes between *V. vinif*era with *V. thunbergi*i or *V. amurensis*. Alcohols and carbonyls were relatively low in all *Vitis*germplasm studied. Terpenoids were abundant in *V. vinifera* with muscat aroma, while esters were dominant in *V. labrusca* and its hybrids with *V. vinifera* or *V. amurensis* [87]. The main aromatic components in three early apple cultivars ("Zaofengtian", "Vista Bella" and "Liaofu") belonging to 12 categories were 1-hexanol, (*E*)-2-hexenal, acetic acid, butyl ester, acetic acid, hexyl ester, *etc.*, representing 93.81% of the total aromatic content. *β*-Damascenone and estragole in "Zaofengtian", which weren't detected in "Vista Bella" or "Liaofu", could be unique components to "Zaofengtian" [88]. In strawberry, differences have been observed between cultivated and wild-type varieties with the monoterpene-linalool and the sesquiterpene nerolidol being the most abundant in cultivated varieties, while oleafinic monoterpenes and myrenyl acetate are more important in the wild-type varieties [89,90].

Insertion of the *rin* gene to reduce ethylene production and slow tomato fruit softening, resulted in some deterioration in flavor quality and reduction in flavor volatiles [91]. Transgenic fruit with antisense amino cyclopropanecarboxylic acid (ACC) synthase had lowered levels of many important flavor volatiles [91]. Fruit with antisense pectinmethylesterase had lowered levels of methanol, while those with downregulated phytoene synthase had lowered levels of carotenoid-derived volatiles [91]. The transformed melon plants displayed a range of decreased alcohol acyltransferase (AAT) activities in the ripening fruit. Therefore, the relative content of volatile esters was reduced, and the average relative content in the transgenic fruit was 78% of the wild-type fruit. However, the relative contents of aldehydes and alcohols increased, the average relative contents were 3.2 times and 2.4 times of those in the wild-type fruit, respectively [92].

3.2. Maturity

Many factors including cultivar, cultural practices, ripeness, harvest maturity and postharvest handling can influence the abundance of volatile compounds in fruit. Of these factors, maturity is one of the critical factors to influence the abundance of volatile compounds in fruit [93]. Ideally, fruit should be harvested at optimal eating quality to optimize volatile content for flavor. However, immature fruit are often harvested in order to increase storage and market life and minimize physical damage and disorder expression. Although immature fruit are more successfully stored and transported, flavor is often lacking due to the close relationship between maturity and volatile biosynthesis [94].

In apples, immature fruits produce low quantities of volatiles at harvest, and lose the capability of volatile production during storage more readily than mature fruit [95]. In "Cigaline" and "Chandler" varieties, C_6aldehydes and alcohols products of the enzymatic breakdown of unsaturated fatty acids, are major contributors to the flavor of immature fruits in the absence of furanones and esters. During fruit ripening, levels of C_6 compounds decrease drastically with increasd furanone acid lactone and ester production [96]. The enhancement of volatile ester production by "Golden Reinders" apples at late maturity stages may have arisen mainly from greater availability of substrates, thus pointing out the relevance of regulating points located upstream of AAT in the pathway [7]. Similar to apples, harvest maturity plays a pivotal role in volatile development of strawberry fruit. C_6 aldehydes were identified as the major compounds in immature white fruit, while furanone and esters are present in three quarters or fully red fruit [96]. Flavor volatile development in melons is also closely linked to fruit maturity with the concentration of total volatile compounds increasing linearly with increasing maturity in cantaloupe melons [97]. El-Mageed noted that ethanol, (*Z*)-3-hexanol and (*E*)-2-hexenal, the most abundant volatiles in whole green and ripe "Fuerte" avocados (*Persea Americana* Mill), declined with ripening, while overripe fruit had the highest ester concentration [98]. Pereira reported that the most abundant volatiles present in the headspace of unripe, diced "Simmonds" avocado were sesquiterpenes and hexanal, whereas the amounts of these compounds was greatly reduced by ripening [99]. During maturation time of avocados cv. "Hass" the concentration of hexanal, (*E*)-2-hexenal and 2,4-hexadienal greatly declined in amount, while acetaldehyde, methyl acetate, pentanal, and β-myrcene were at higher concentrations in mature fruit and may also have contributed to the overall flavor [100]. In detached "Pluk Mai Lie" papaya fruit, 2-ethyl-

1-hexanol was found specifically in green fruit, ethyl octanoate emerged only in fully-ripe fruit. Furthermore, benzyl isothiocyanate was the most abundant volatile present in fruit at every stage except full ripening. The levels of methanol and ethanol sources in fruit increased steadily throughout ripening, with esters formed from ethyl alcohol predominating from the half-ripe through the senescence phases. The alcohol dehydrogenase (ADH) activity in the mesocarp increased dramatically during the early ripening stages, whereas ATT was active throughout ripening [101]. Yang *et al.* [102] reported that all the alcohols and carbonyls, along with most of the C_6 compounds and terpenoids, were evident before veraison in three different flavour table-grapes, "Jingxiu", "Bimeijia" and "Jingya", while most of the esters were detected at or after veraison. C_6compounds increased in the early period of maturation, and then decreased. Most alcohols and carbonyls tended to continuously decrease during ripening. Except for geraniol, terpenoids increased until maturation, then decreased. Some esters continued to increase after maturation. Principal component analysis showed that terpenoids and esters were the characteristic volatiles of ripe "Bimeijia" and "Jingya" grapes, respectively. "Bimeijia" had the highest terpenoid content at maturity, while "Jingya" continued to accumulate some esters after maturation.

3.3. Pre-Harvest Factors

Pre-harvest factors such as sunlight, water availability, fertilization, and chemical applications affect crop growth, and can affect internal quality characteristics of the harvested product, including flavor. Heavy rains prior to harvest dilute flavor compounds in tomatoes (*Solanum lycopersicum* L.). Grape aroma potential was highest in vines under mild water deficit and moderate nitrogen supply. Severe water deficit stress seemed to limit aroma potential, as did nitrogen deficiency [103]. Volatile production of "Golden Delicious" apples was affected by aminoethoxyvinylglycine (AVG) application, especially in the case of esters and alcohols, which presented 59% and 33% lower values in AVG-treated apples at the end of storage. A higher concentration of aldehydes was observed at harvest time in AVG-treated apples. Higher correlations between ethylene production and aldehydes and alcohols were found in control apples than in AVG-treated apples. However, AVG treatment negatively affects the production of some volatile compounds [104].

3.4. Postharvest Handling

Various techniques are used to extend the shelf-life of fruits after harvest. These storage techniques and treatments involve cold, heat, irradiation, different storage atmospheres, and chemical applications. These postharvest handlings can also affect the aroma components and concentrations.

3.4.1. Temperature

Storage temperature is a fundamental factor affecting the flavor of fruits. Changes in storage temperatures had no major effects on aroma volatile contents in chilling-tolerant "*Or*" mandarins; however, in chilling-sensitive "Odem" mandarins, storage at 2 °C caused accumulation of 13 volatiles, mainly terpenes and their derivates, whereas storage at 8 °C resulted in decreases of six volatiles, comprising five terpenes and one terpene derivative [105]. Production of volatiles was markedly influenced by storage temperature and time in melting flesh peach "Hujingmilu" fruit. In general, fruit at 5 °C were sensitive to chilling injury and had the lowest levels of volatile compounds, especially fruity note volatiles such as esters and lactones [67]. Refrigeration of tomato induced changes in levels of 3-methylbutanal, linalool, guiacol, hexanol, *trans*-2-hexenal and *trans*-3-hexenol, and some of these alterations may be explained by a decrease in ADH enzyme activity [106]. During cold storage at 4 °C, acetate esters declined and non-acetate esters increased in fresh-cut cantaloupe and honeydew melons. A significant shift in the ratio between acetate and non-acetate esters can be seen even after 2 days in storage. The recycling of esters during storage of fresh-cut melons may lead to an imbalance of characteristic volatiles [107]. Both chilling and heating of tomatoes reduced C_6 aldehyde and alcohol aroma volatiles immediately after treatment, and the levels of aldehydes did not fully recover after 4 days at 20 °C. Chilling-induced inhibition of C_6 volatile production may be due to down-regulation of gene expression, and subsequent reduction of hydroperoxide (HPL) and ADH enzyme activities in the oxylipin pathway. Heating-inhibition of C_6 volatile production, however, does not appear to be due to down-regulation of gene expression, but HPL and ADH activities were briefly suppressed [108].

3.4.2. Storage Atmosphere

Lowering O_2 and raising CO_2 can maintain the quality of many fresh fruits for extended periods. However, exposure of fresh product to O_2 levels below their tolerance level can increase anaerobic respiration and lead to the development of off-flavor. Storage of fruit under controlled atmosphere (CA) conditions can reduce the capacity of several fruit to produce ethylene and alter production of aroma volatiles [109]. Storage of peach fruit cv "Rich Lady" for 15 days, under 3% O_2 + 10% CO_2 at 2 °C, improved juiciness, sweetness, perception of flavor, emission of aroma volatile compounds and sensory acceptance in comparison with fruit stored in cold air [110]. Decreasing O_2 levels and increasing CO_2 levels of "Titania" blackcurrants retarded the capacity of 3-week stored fruit to synthesize terpenes. Differential changes among the various groups of terpenes were more important, where terpene alcohols reached a peak in 6-week air-stored fruit, and storing berries under a high CO_2 level (18 kPa) and/or decreasing O_2 level (2 kPa) resulted in lower biosynthesis of these alcohols compared to control fruit. CA storage conditions led to a transitory reduction in the emission of alcohols but a recovery was recorded with prolonged storage. Non-terpene esters differed greatly in storage, in particular the ester ethyl butanoate. Air-stored fruit at both sampling dates synthesized significantly higher amounts of esters than freshly harvested fruit but a significant decline was observed for branched butyl substances (2-methyl butyl butanoate) after 6 weeks storage [111].

Use of packaging and edible coatings can create a modified atmosphere (MA) with reduced O_2 and elevated CO_2 levels, similar to that of CA. Use of edible coatings affects flavor and the level of volatile flavor compounds in citrus, apple and mango fruit [112,113,114]. The coating barrier probably induced anaerobic respiration and the synthesis of ethanol and acetaldehyde, and entrapped volatiles, including ethanol and acetaldehyde [115]. The effects of different edible coatings on mango fruit showed that mango "carnauba" was effective in retarding fruit ripening, retaining fruit firmness, and improving fruit quality attributes including levels of fatty acids and aroma volatiles. Semperfresh and *A. vera* gel (1:1 or 100%) slightly delayed fruit ripening but reduced fruit aroma volatile development. *A. vera* gel coating did not exceed the commercial mango "carnauba" and Semperfresh in retarding fruit ripening and improving aroma volatile biosynthesis [113].

3.4.3. Chemical Application

In addition to CA, other gaseous treatments of fruits and vegetable have been reported. Use of ethylene to synchronize ripening has been practiced for years on banana and tomato, and for degreening of citrus. Ethylene treatment of tomato fruit alters volatile levels [116]. 1-methylcyclopropene (1-MCP) treatment of peach fruit "Tardibelle" altered the supply of alcohol and acyl-CoA precursors, leading to significant changes in the emission of some volatile esters, particularly of the straight-chain type [8].

Other chemical treatments of fresh product may also affect flavor. Calcium treatment of fruit is a widely used practice aimed mainly at avoiding the development of bitter pit. Calcium treatment of commercially mature "Golden Reinders" apples notably enhanced the production of aroma volatile compounds after mid-term storage under air and, to a lesser extent, under standard CA. Aroma volatile production was severely depleted in ultra-low oxygen atmosphere stored samples, and calcium treatment could not overcome this inhibition [117]. Methyl jasmonate (MJ) alone and in conjunction with ethanol, is able to modify the biochemical pathways of volatile compounds [118]. Reports in the literature have described the impact of MJ on biosynthesis of volatile compounds in both climacteric fruit, such as apples and non-climateric fruit, such as strawberries

[119,120,121,122]. The application of MJ may avoid undesirable alterations in the volatile fraction occurring postharvest and during storage, as well as minimizing aroma losses. Postharvest MJ treatments in combination with ethanol on the formation of aroma constituents in berryfruit (raspberries, strawberries and blackberries) showed different effects according to berry species. In contrast to raspberries, which exhibited a significant decline in the total amount of volatiles after treatment, a significant enhancement of total volatile compounds was observed in strawberries, while no significant effect was found in blackberries. Esters and terpene compounds responded similarly in strawberries and blackberries suggesting similarity in the biochemistry of their aroma synthesis. In contrast, raspberry volatile compounds showed a different pattern, reflecting different biosynthetic pathways for aroma formation in raspberry. The natural volatile compounds, MJ and ethanol, seemed to have either promoting effects on the formation of the (–)-enantiomers of chiral terpenes and ionones or inhibitory effects on the synthesis of the corresponding (+)-enantiomers [123].

4. VOLATILE AROMA COMPOUNDS BIOSYNTHESTIC PATHWAYS AND RELATED ENZYMES

As volatiles are comprised of at least five chemical classes, there are several pathways involved in volatile biosynthesis. These have not been fully described but appear to be common for different fruits. Volatiles important for aroma and flavor are biosynthesized from amino acids, membrane lipids and carbohydrates [4]. Although volatile compounds are synthesized via a few major biochemical pathways, various forms of enzymatic modifications such as hydroxylations, acetylations, and methylations, add to the diversity of emitted volatiles by increasing their volatility at the final step of their formation [124,125]. An important step in the biosynthetic pathway of aroma compounds is the availability of primary precursor substrates, including fatty acids and amino acids, which are highly regulated during fruit development in terms of amount and composition [126].

4.1. Fatty Acids Pathway

Fatty acids are major precursors of aroma volatiles in most fruit [4]. Fatty acid-derived straight-chain alcohols, aldehydes, ketones, acids, esters and lactones ranging from C_1 to C_{20} are important character-impact aroma compounds that are responsible for fresh fruit flavors with high concentrations, and are basically formed by three processes: α-oxidation, β-oxidation and the lipoxygenase pathway [127]. Aroma volatiles in intact fruit are formed via the β-oxidation biosynthetic pathway, whereas when fruit tissue is disrupted, volatiles are formed via the lipoxygenase (LOX) pathway [128]. Nevertheless, some studies suggest that increasing availability of fatty acid, along with higher membrane permeability, during fruit ripening might allow the LOX pathway to become active in intact plant tissue and to function as an alternative to β-oxidation [30]. Many of the aliphatic esters, alcohols, acids, and carbonyls found in fruits are derived from the oxidative degradation of linoleic and linolenic acids [16]. In addition, some of the volatile compounds derived from enzyme-catalyzed oxidative breakdown of unsaturated fatty acids may also be produce by autoxidation [129]. Autoxidation of linoleic acid produces 9,13-hydroperoxides, whereas linolenic acid also produces 12,16-hydroperoxides [27]. Hexanal and 2,4-decadienal are the primary oxidation products of linoleic acid, while autoxidation of linolenic acid produces 2,4-heptadienal as the major product. Further autoxidation of these aldehydes leads to the formation of other volatile products [129].

4.1.1. β-Oxidation

Although the degradation of straight chain fatty acids by α and β-oxidation is a major process for the formation of flavor molecules in all organisms, the specific pathways in plants are not well understood. Baker *et al.* described varied roles for this pathway in relation, not only to fatty acid catabolism but also, to amino acid metabolism and biosynthesis of hormonal compounds [130]. β-Oxidation results in successive removal of C_2 units (acetyl CoA) from the parent fatty acid. The detailed mechanisms of conventional β-

oxidation are well established [131]. Sanz *et al.* reported that β-oxidation of fatty acids is the primary biosynthetic process providing alcohols and acyl coenzyme A (CoAs) for ester formation [4]. Fatty acid acyl-CoA derivatives are converted to shorter chain acyl CoAs by losing two carbons in every round of the *β*-oxidation cycle, requiring flavinadenine dinucleotide (FAD), nicotinamide adenine dinucleotide (NAD), and free CoA. Acyl CoAs are reduced by acyl CoA reductase to aldehyde that in turn is reduced by ADH to alcohol for use by AAT to produce esters [132]. Pear and apple aromas have been two classical examples of volatile formation through the *β*-oxidation pathway [133].

The biosynthesis of lactones, key aroma components in fruits such as peach and nectarine (γ-decalactone and γ-dodecalactone), pineapple (δ-octalactone), or coconut (*Cocos nucifera* L.) (γ-octalatcone), is also associated with the β-oxidation pathway [134]. In fact, most hypotheses on lactone biosynthesis in fruits put in contact the two major pathways producing aroma compounds from fatty acid, *β*-oxidation, and LOX [4]. Despite the importance of these compounds in fruit aroma, there is a lack of enzymatic studies in fruits, and microorganisms serve as a model for studying lactone biosynthesis [135].

4.1.2. Lipoxygenase (LOX)

The metabolism of polyunsaturated fatty acids, via the first LOX-catalyzed step and the subsequent reactions, is commonly known as the LOX pathway. Saturated and unsaturated volatile C_6 and C_9 aldehydes and alcohols are important contributors to the characteristic flavors of fruits, vegetables and green leaves. The short-chain aldehydes and alcohols are produced by plants in response to wounding and play an important role in the plants defense strategies and pest resistance [136,137]. At least four enzymes are involved in the biosynthetic pathway leading to their formation: LOX, HPL, 3(Z), 2(*E*)-enal isomerase and ADH. When fruit are homogenised, linoleic and linolenic acid are oxidised to various C_6 and C_9 aldehydes [138]. In intact fruit, enzymes in the LOX pathway and their substrates have different subcellular locations, preventing formation of volatile compounds [4]. During ripening, cell walls and membranes may become more permeable, allowing the LOX pathway to become active without tissue disruption [4]. The LOX biosynthetic pathway has the potential to provide substrates for ester production [139]. If the LOX biosynthetic pathway were active during ripening, it would act as an alternative to β-oxidation of fatty acids.

As shown in Scheme 1, volatile fatty acid derivatives such as *trans*-2-hexenal, *cis*-3-hexenol and methyl jasmonate are derived from C_{18} unsaturated fatty acids including linoleic acid or linolenic acid, which undergo dioxygenation in a reaction catalyzed by LOX [140]. These enzymes can catalyze the oxygenation of polyenoic fatty acids at C_9 or C_{13} positions yielding two groups of compounds, the 9-hydroperoxy and the 13-hydroperoxy derivatives of polyenoic fatty acids. These derivatives can be further metabolized by an array of enzymes, including allene oxide synthase (AOS) and HPL, which represent two branches of the lipoxygenase pathway yielding volatile compounds. In the AOS branch of the lipoxygenase pathway, 13-hydroxyperoxy linolenic acid is converted to 12,13-epoxyoctadecatrienoic acid by AOS [140]. A series of subsequent enzymatic reactions leads to the formation of jasmonic acid, which can in turn be converted to the volatile ester, methyl jasmonate, by the enzyme jasmonic acid carboxyl methyltransferase [141]. In the HPL branch of the LOX pathway, the oxidative cleavage of hydroperoxy fatty acids through the action of HPL leads to the formation of short chain C_6 or C_9 volatile aldehydes (e.g., 3-hexenal or 3,6-nonadienal) and the corresponding C_{12} or $C_9\omega$-fatty acids (e.g., 12-oxo-dodecenoic acid or 9-oxononanoic acid). HPL C_6aldehyde products can be further converted to their isomers by spontaneous rearrangement by alkenal isomerases, or reduced to alcohols by the action of ADH [142].

4.2. Amino Acid Pathway

Amino acid, such as alanine, valine, leucine, isoleucine, phenylalanine and aspartic acid, are also involved in aroma biosynthesis in fruit as direct precursors, and their metabolism is responsible for the production of a broad number of compounds, including alcohols, carbonyls, acids and esters [4,143]. Most of the information available to date on the biosynthesis of amino acid-derived volatiles in plants is based on precursor feeding experiments with radio-labeled, stable-isotope-labeled, or unlabeled precursors. The general scheme of biosynthesis is thought to proceed in a similar way as that in bacteria or yeast, where these pathways have been studied more extensively [144,145]. Amino acids can undergo an initial deamination or transamination leading to the formation of the corresponding α-keto acid (Scheme 2). Subsequent decarboxylation followed by reductions, oxidations and/or esterifications give rise to aldehydes, acids, alcohols and esters [16]. Branched chain volatile alcohols, aldehydes and esters in fruits such as banana, apple, strawberry and tomato arise from the branched chain amino acids leucine, isoleucine and valine [10,146,147,148]. These amino acids can also be the precursors of acyl-CoAs, which are used in alcohol esterification reactions catalyzed by AATs. Indeed, isoleucine could give rise to 3-methylbutanol and 2-methylbutyryl-CoA, both used in an esterification reaction to yield the ester 3-methylbutyl 2-methylbutanoate in banana [149]. Methionine could be the precursor of sulphur-containing volatiles such as dimethyldisulfide and volatile thioesters [18]. In strawberry, it has been suggested that alanine serves as a precursor for volatile ethyl esters, which can be produced by AAT [149,150].

4.3. Terpenoids Pathway

The terpenoids compose the largest class of plant secondary metabolites with many volatile representatives. Hemiterpenes (C_5), monoterpenes (C_{10}), sesquiterpenes (C_{15}), homoterpenes (C_{11} and C_{16}), and some diterpenes (C_{20}) have a high vapor pressure allowing their release into the atmosphere. As shown in Scheme 3, terpenoids are derived from the universal C_5 precursor isopentenyl diphosphate (IPP) and its allylic isomer dimethylallyl diphosphate (DMAPP),

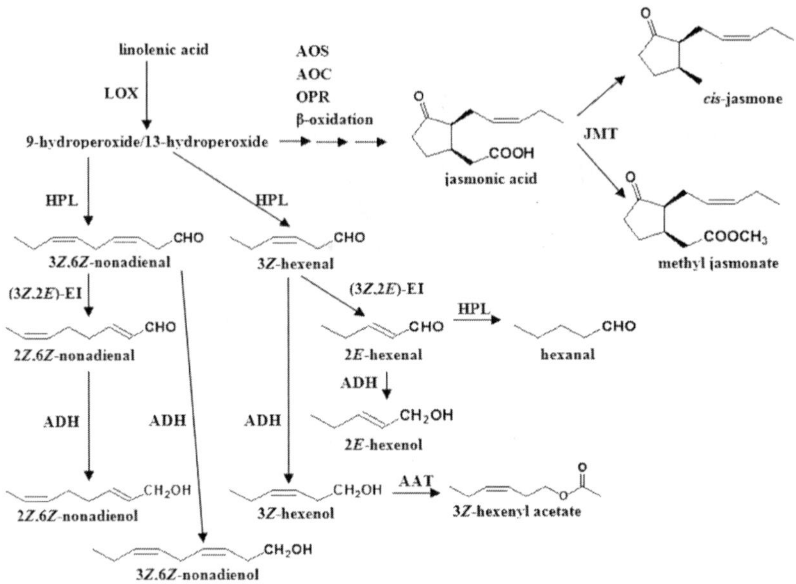

Scheme 1. Linolenic acid-derived flavor molecules.AAT, alcohol acyl CoA transferase; ADH, alcohol dehydrogenase; AER, alkenal oxidoreductase; AOC, allene oxide cyclase; AOS, allene oxide synthase; HPL, hydroperoxide lyase; JMT, jasmonate methyltransferase; LOX, lipoxygenase; OPR, 12-oxo-phytodienoic acid reductase; 3*Z*,2*E*-EI, 3*Z*,2*E*-enal isomerase.

which in higher plants are generated from two independent pathways located in separate intracellular compartments. In cytosol, IPP is derived from the long-known mevalonic acid (MVA) pathway that starts with the condensation of acetyl-CoA [151]. In plastids, IPP is formed from a MVA-independent pathway (or MEP pathway) with pyruvate and glyceraldehydes 3-phosphate as direct precursors and methylerythritol phosphate (MEP) as the key intermediate [152]. Initial research indicated that the cytosolic IPP serves as a precursor of farnesyl diphosphate (FPP) for sesquiterpenes and triterpenes, whereas the plastidial IPP provides the precursors for geranyl diphosphate (GPP) and GGPP for mono-, di-, and tetra-terpenes. However, cross-talk between these two IPP biosynthetic pathways is prevalent, particularly in the direction from plastids to cytosol [153,154].

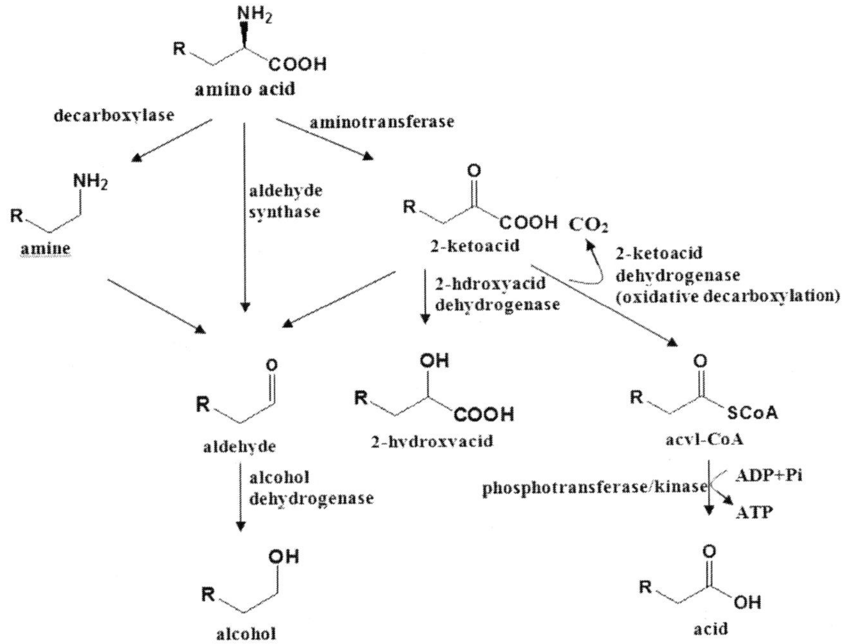

Scheme 2. Biosynthetic routes for amino acid degradation to volatiles in plants and microorganisms.

In plastids, DMAPP generated from the MEP pathway is used by isoprene synthases for isoprene formation [155,156]. In the cytosol and in plastids, IPP and DMAPP are used by prenyltransferases to produce prenyl diphosphates. In cytosol, the condensation of two molecules of IPP and one molecule of DMAPP catalyzed by the enzyme farnesyl pyrophosphate synthase (FPPS) results in the formation of FPP (C_{15}), the natural precursor of sesquiterpenes [157]. In plastids, a head-to-tail condensation of one molecule of IPP and one molecule of DMAPP catalyzed by geranyl pyrophosphate synthase (GPPS) forms GPP (C_{10}), the universal precursor of all the monoterpenes [158]. The condensation of one molecule of DMAPP with three molecules of IPP by the action of geranylgeranyl pyrophosphate synthase (GGPPS) yields GGPP, the C_{20} diphosphate precursor of diterpenes [158,159]. Following the formation of the acyclic precursors GPP, FPP, and GGPP, a wide range of structurally diverse cyclic and acyclic monoterpenes, sesquiterpenes, and diterpenes is generated through the action of a large family of enzymes known as terpene synthases/cyclases (TPSs) [160,161]. One of the most outstanding properties of these enzymes is their proclivity for making multiple products from a single prenyl diphosphate substrate [162].

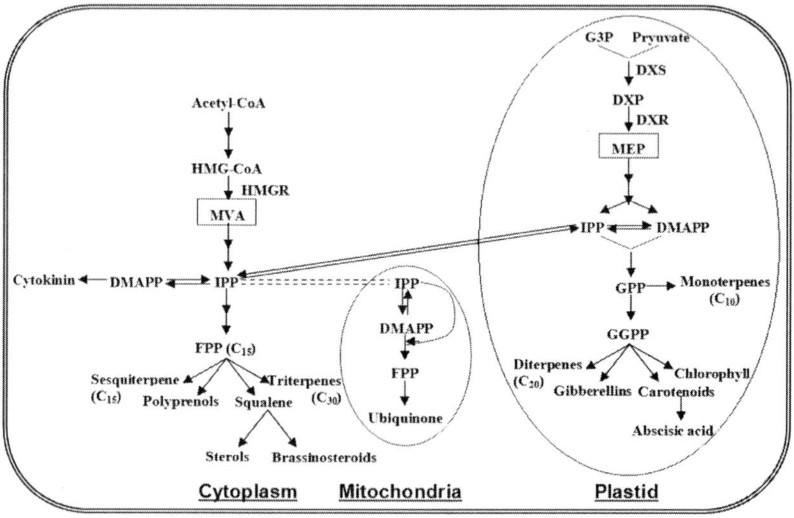

Scheme 3. The biosynthesis pathway of isoprenoids in plant cell. DMAPP, dimethylallyl diphosphate; DXP, 1-deoxy-D-xylulose-5-phosphate; DXS, DXP synthase; DXR, DXP reductoisomerase; FPP, farnesyl diphosphate; G3P, glyceraldehyde 3-phosphate; GPP, geranyl diphosphate; GGPP, geranyl geranyl diphosphate; HMGR, HMG-CoA reductase; IPP, isopentenyl diphosphate; MEP, methylerythritol phosphate; MVA, mevalonic acid.

Many of the terpene volatiles are direct products of terpene synthases, while others are formed through alterations of the primary terpene skeletons made by TPSs by hydroxylation, dehydrogenation, acylation, and other reaction types [124]. 3-Hydroxylation of limonene by a P450 enzyme yields *trans*-isopiperitenol, a volatile compound found in mint (*Mentha aquatic*) [163], while 6-

hydroxylation of limonene by another P450 enzyme results in the formation of *trans*-carveol, which undergoes further oxidation by nonspecific dehydrogenases to form carvone, a major aroma volatile of caraway fruits (*Carum carvi* L.) [164,165]. Another monoterpene alcohol, geraniol, was also found to be converted to the corresponding aldehyde by dehydrogenases in the glands of sweet basil [166]. On the other hand, an acetylation of geraniol by acetyltransferase generates geranyl acetate, a volatile compound found in the scent of many plant species [167,168]. Recently, reduction of geraniol to *S*-citronellol, the precursor of the potent odorant rose oxide, was confirmed enzymatically in grape mesocarp [169]. Modification reactions are also involved in the formation of terpenoids with irregular acyclic C_{16} and C_{11} carbon skeletons, so called homoterpenes, which are mainly emitted from injured tissues. Although the exact biosynthetic routes to homoterpenes, (*E,E*)-TMTT (C_{16}) and (*E*)-DMNT (C_{11}) still remain unclear, it is believed that they are derived from geranyl-linalool (C_{20}) and (3*S*)-(*E*)-nerolidol (C_{15}), respectively, by oxidative degradation possibly catalyzed by cytochrome P450 enzymes [170].

4.4. Carotenoid Pathway

Carotenoid and their apocarotenoid derivatives are isoprenoid molecules important for the primary and secondary metabolisms of plants and other living organisms [171]. The oxidative cleavage of carotenoids leads to the production of apocarotenoids and is catalyzed by a family of carotenoid cleavage dioxygenases (CCDs) [172]. Apocarotenoid volatiles are synthesized only at the latest stage of ripening, even though the CCD enzymes are present throughout fruit development [173]. CCDs often exhibit substrate promiscuity, which probably contributes to the diversity of apocarotenoids found in nature [172]. CCDs also differ for their subcellular localization: some, like CCD1, are predicted to be cytosolic; others possess transit peptides for plastid or plastoglobule targeting [174]. In general, the biosynthesis of carotenoid-derived volatile compounds occurs via three steps: an initial dioxygenase cleavage yielding apocarotenoids, followed by enzymatic transformations of these apocarotenoids leading to the formation of polar aroma precursors, and finally acid-catalyzed conversions of these precursors to volatile compounds [22]. However, in some cases a volatile product is the result of the initial dioxygenase cleavage step, as was shown for *β*-ionone in Arabidopsis (*Arabidopsis thaliana*) [175], tomato (*Solanum lycopersicum*) and petunia (*Petunia hybridea*) [172,176].

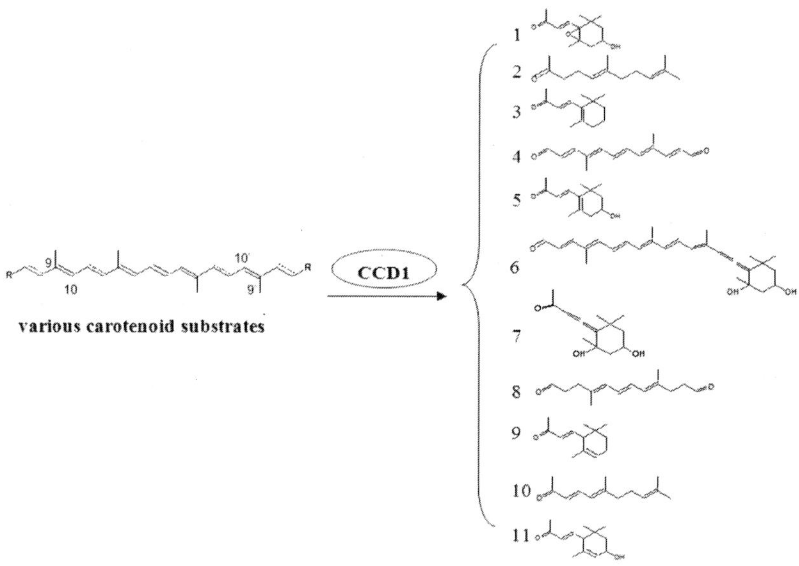

Scheme 4. Carotenoids and their degradation products. 1, 5,6-epoxy-3hydroxy-*β*-ionone; 2, geranayl acetone (6,10-dimethyl-5,9-undecadien-2-one); 3, *β*-ionone; 4, C_{14} dialdehyde (4,9-dimethyldodeca-2,4,6,8,10-pentaene-1,12-dial); 5, 3-hydroxy-*β*-ionone; 6, the cleavage product of 9-*cis*-neoxanthin on the 9,10 double bond; 7, the cleavage product of 9-*cis*-neoxanthin on the 9′,10′ double bond; 8, 4,9-dimethyldodeca-4,6,8-trienedial; 9, *α*-ionone; 10, pseudoionone (6,10-dimethyl-3,5,9-undecatrien-2-one); 11, 3-hydroxy-α-ionone.

CCDs exhibit specificity for the double bond that they cleave but many are promiscuous in their substrate choice. In plants, CCD1 and CCD7 cleave the 9,10 double bonds of their respective carotenoid substrates,*At*CCD1 cleaves linear and cyclic carotenoids at the 9,10 and 9′,10′ positions. For example, when β-carotene serves as the substrate, *At*CCD1 produces two C_{13} products (both β-ionone) and a central C_{14} dialdehyde (Scheme 4). CCD7 can cleave both linear and cyclic carotenoid substrates at the 9,10 double bond [177]. Schwartz *et al.* determined that CCD7 cleaves β-carotene asymmetrically, generating a C_{13} ketone (*β*-ionone) and a C_{27} aldehyde (10`-apo-β-carotenal). CCD8 can then cleave the C_{27} aldehyde at its 13,14 double bond [178]. Subsequent data demonstrate, however, that CCD8 can act directly upon carotenoid substrates [179]. In addition, *Bo*LCD from *Bixa orellana* was reported to cleave at the lycopene 5,6 double bond [179], and*Cs*ZCD from *Crocus sativus* was found to cleave at the 7,8 double bond of zeaxanthin [180].

5. CONCLUSIONS

Fruit aroma is an important indicator to reflects the quality of fruit flavor. Fruits synthesize and emit a large variety of aroma volatile compounds with terpenoids and fatty-acid derivatives the dominant classes. Whereas some volatiles are probably common to almost all fruits, others are specific to only one or a few related fruits. Production of volatiles was markedly influenced by many factors, to date we have a limited understanding of how these factors interact to determine the actual volatile composition and resulting flavor of the fruit. In general, more than one biochemical pathway is responsible for a blend of volatile compounds released from different fruits. Within this final issue, further information about the genes that are involved in the synthesis of aroma volatile should be reviewed.

ACKNOWLEDGMENTS

This work was financially supported by the National Natural Science Foundation of China (Grant no. 31171934), the Priority Academic Program Development from Jiangsu Government and the Science & Technology Innovation Foundation of Yangzhou University (Grant no. 2012CXJ062). The authors are grateful to Prof. Osman Ali Sidahmed (University of Gezira, Sudan) for assistance in reading through the manuscript and making appropriate amendment.

CONFLICTS OF INTEREST

The authors declare no conflict of interest.

REFERENCES

1. Ogundiwin, E.A.; Peace, C.P.; Gradziel, T.M.; Parfitt, D.E.; Bliss, F.A.I.; Crisosto, C.H.A. Fruit quality gene map of Prunus. BMC Genomics 2009, 10, 587.

2. Maffei, M.E. Changes in biosynthesis of aroma volatile compounds during on-tree maturation of "Pink Lady" apples. South Afr. J. Bot. 2010, 76, 612–631.

3. Cheong, K.W.; Tan, C.P.; Mirhosseini, H.; Hamid, N.S.A.; Osman, A.; Basri, M. Equilibrium headspace analysis of volatile flavour compounds extracted from soursop (Anoan muricata) using solid phase microextraction. Food Res. Int. 2010, 43, 1267–1276.

4. Sanz, C.; Olias, J.M.; Perez, A.G. Aroma biochemistry of fruits and vegetables. In Phytochemistry of Fruit and Vegetables; Oxford University Press Inc.: New York, NY, USA, 1997; pp. 125–155.

5. Schwab, W.; Davidovich-Rikanati, R.; Lewinsohn, E. Biosynthesis of plant-derived flavor compounds.Plant J. 2008, 54, 712–732.

6. Tucker, G.A. Introduction. In Biochemistry of Fruit Ripening; Seymour, G.B., Taylo, R.J.E., Tucker, G.A., Eds.; Chapman & Hall: London, UK, 1993; pp. 1–51.

7. Ortiz, A.; Graell, J.; Lara, I. Volatile ester-synthesising capacity throughout on-tree maturation of "Golden Reinders" apples. Sci. Hortic. 2011, 131, 6–14.

8. Ortiz, A.; Graell, J.; López, M.L.; Echeverría, G.; Lara, I. Volatile ester-synthesising capacity in "Tardibelle" peach fruit in response to controlled atmosphere and 1-MCP treatment. Food Chem. 2010,123, 698–704.

9. Song, J.; Forney, C.F. Flavour volatile production and regulation in fruit. Can. J. Plant Sci. 2008, 88, 537–550.

10. Goff, S.A.; Klee, H.J. Plant volatile compounds: Sensory cues for health and nutritional value. Science2006, 311, 815–819.

11. Drawert, F.; Heimann, W.; Emberger, R.; Tressl, R. Über die Biogenese von Aromastoffen bei Pflanzen und Frü chten. IV. Mitt Bildung der Aromamstoffe des Apfels im Verlauf des Wachstums und bei der Largerung. Zeit. Lebens. Unters. Forsch 1969, 140, 65–87.

12. Berger, R.G.; Drawert, F.; Kollmannsberger, H. Geruchsaktive spurenkomponenten des bananen aromas. Chem. Mikrobiol. Technol. Lebensm. 1986, 10, 120–124.

13. Bruckner, B. Fruit and Vegetable Flavour: Recent Advances and Future Prospects; Woodhead Publishing Limited: Abington Hall, Cambridge, UK, 2008.

14. Buttery, R.G. Quantitative and sensory aspects of flavor of tomato and other vegetable and fruits. InFlavor Science: Sensible Principles and Techniques; Acree, T.E., Teranishi, R., Eds.; ACS: Washington, DC, USA, 1993; pp. 259–286.

15. Sarry, J.E.; Gunata, Z. Plant and microbial glycoside hydrolases: Volatile release from glycosidic aroma precursors. Food Chem. 2004, 87, 509–521.

16. Reineccius, G. Flavor Chemistry and Technology, 2nd ed.; CRC Press: Boca Raton, FL, USA, 2006.

17. Carcia, C.V.; Stevenson, R.J.; Atkinson, R.G.; Winz, R.A.; Yong Quik, S. Changes in the bound aroma profiles of "Hayward" and "Hort16A" kiwifruit (Actinidia spp.) during ripening and GC-olfactometry analysis. Food Chem. 2013, 137, 45–54.

18. Wyllie, S.G.; Leach, D.N.; Wang, Y.; Shewfelt, R.L. Key aroma compounds in melons: Their development and cultivar dependence. In Fruit Flavors: Biogenesis, Characterization, and Authentication; Rouseff, R.L., Leahy, M.M., Eds.; American Chemical Society: Washington, DC, USA, 1995; pp. 248–257.

19. Zabetakis, I.; Holden, M.A. Strawberry flavour: Analysis and biochemistry. J. Sci. Food Agric. 1997, 74, 421–434.

20. Ninio, R.; Lewinsohn, E.; Mizrahi, Y.; Sitrit, Y. Quality attributes of storage koubo (Cereus peruvianus(L.) Miller) fruit. Postharvest Biol. Technol. 2003, 30, 273–280.

21. Akakabe, Y.; Sakamoto, M.; Ikeda, Y.; Tanaka, M. Identification and characterization of volatile components of Japanese sour citrus fruit Citrus nagato-yuzukichi Tanaka. Biosci. Biotechnol. Biochem.2008, 72, 1965–1968.

22. Winterhalter, P.; Rouseff, R.L. Carotenoid-derived aroma compounds: An introduction. In Carotenoid-Derived Aroma Compounds; Winterhalter, P., Rouseff, R.L., Eds.; American Chemical Society: Washington, DC, USA, 2002; pp. 1–17.

23. Nijssen, L.M.; van Ingen-Visscher, C.A.; Donders, J.J.H. VCF Volatile Compounds in Food: database (Version 13.1.). Zeist (The Netherlands): TNO Triskelion Recuperato da, 2011.

24. Dixon, J.; Hewett, E.W. Factors affecting apple aroma/flavour volatile concentration: A review. N.Z.J. Crop Hortic. Sci. 2000, 28, 155–173.

25. Rizzolo, A.; Grassi, M.; Zerbini, P.E. Influence of harvest date on ripening and volatile compounds in the scab-resistant apple

cultivar "Golden Orange". J. Hortic. Sci. Biotech. 2006, 81, 681–690.

26. Holland, D.; Larkov, O.; Bar-Yaákov, I.; Bar, E.; Zax, A.; Brandeis, E. Developmental and varietal differences in volatile ester formation and acetyl-CoA: Alcohol acetyl transferase activities in apple (Malus domestica Borkh.) fruit. J. Agric. Food Chem. 2005, 53, 7198–7203.

27. Berger, R.G. Flavours and Fragrances-Chemistry, Bioprocessing and Sustainability; Springer-Verlag: Berlin, Germany, 2007.

28. Villatoro, C.; Altisent, R.; Echeverria, G.; Graell, J.; Lopez, M.L.; Lara, I. Changes in biosynthesis of aroma volatile compounds during on-tree maturation of "Pink Lady" apples. Postharvest Biol.Technol.2008, 47, 286–295.

29. Rudell, D.R.; Mattinson, D.S.; Mattheis, J.P.; Wyllie, S.G.; Fellman, J.K. Investigations of aroma volatile biosynthesis under anoxic conditions and in different tissues of "Redchief Dilicious" apple fruit (Malus domestica Borkh.). J. Agric. Food Chem. 2002, 50, 2627–2632.

30. Guadagni, D.G.; Bomben, J.L.; Hudson, J.S. Factors influencing the development of aroma in apple peel. J. Sci. Food Agric. 1971, 22, 110–115.

31. Defilippi, B.G.; Dandekar, A.M.; Kader, A.A. Relationship of ethylene biosynthesis to volatile production, related enzymes and precursor availability in apple peel and flesh tissues. J. Agric. Food Chem. 2005, 53, 3133–3141.

32. Rapparini, F.; Predieri, S. Pear fruit volatiles. In Horticultural Reviews; Janick, J., Ed.; John Wiley & Sons: Hoboken, NJ, USA, 2003; pp. 237–324.

33. Kahle, K.; Preston, C.; Richling, E.; Heckel, F.; Schreier, P. On-line gas chromatography combustion/pyrolysis isotope ratio mass spectrometry (HRGC-C/P-IRMS) of major volatiles from pear fruit (Pyrus communis) and pear products. Food Chem. 2005, 91, 449–455.

34. Rizzolo, A.; Sodi, C.; Polesello, A. Influence of ethylene removal on the volatile development in Passa crassana pears stored in a controlled atmosphere. Food Chem. 1991, 42, 275–285.

35. Rizzolo, A.; Cambiaghi, P.; Grassi, M.; Zerbini, P.E. Influence of 1-methylcyclopropene and storage atmosphere on changes in volatile compounds and fruit quality of conference pears. J. Agric. Food Chem. 2005, 53, 9781–9789.

36. Obando-Ulloa, J.M.; Moreno, E.; Garcıa-Mas, J.; Nicolai, B.; Lammertync, J.; Monforte, J.A.; Fernandez-Trujillo, J.P. Climacteric or non-climacteric behavior in melon fruit 1. Aroma volatiles. Postharvest Biol. Technol. 2008, 49, 27–37.

37. Perry, P.L.; Wang, Y.; Lin, J.M. Analysis of honeydew melon (Cucumis melo var. Inodorus) flavor and GC/MS identification of (E,Z)-2,6-nonadienyl acetate. Flav. Frag. J. 2009, 24, 341–347.

38. Portnoy, V.; Benyamini, Y.; Bar, E. The molecular and biochemical basis for varietal variation in sesquiterpene content in melon (Cucumis melo L.) rinds. Plant Mol. Biol. 2008, 66, 647–661.

39. Aubert, C.; Bourger, N. Investigation of volatiles in Charentais cantaloupe melons (Cucumis melo var. cantalupensis). Characterization of aroma constituents in some cultivars. J. Agric. Food Chem. 2004, 52, 4522–4528.

40. Aubert, C.; Pitrat, M. Volatile compounds in the skin and pulp of Queen Anne's pocket melon. J. Agric. Food Chem. 2006, 54, 177–8182.

41. Ibdah, M.; Azulay, Y.; Portnoy, V.; Wasserman, B.; Bar, E.; Meir, A.; Burger, Y.; Hirchberg, J.; Schaffer, A.A.; Katzir, N.; Tadmor, Y.; Lewinsohn, E. Functional characterization of CmCDD1, a carotenoid cleavage dioxygenase from melon. Phytochemistry 2006, 67, 1579–1589.

42. Manriquez, D.; El-Sharkawy, I.; Flores, F.B.; El-Yahyaoui, F.; Regad, F.; Bouzayen, M.; Latche, A.; Pech, J.C. Two highly divergent alcohol dehydrogenases of melon exhibit fruit

ripening-specific expression and distinct biochemical characteristics. Plant Mol. Biol. 2006, 61, 675–685.

43. Burger, Y.; Sáar, U.; Paris, H.S.; Lewinsohn, E.; Katzir, N.; Tadmor, Y.; Schaffer, A.A. Genetic variability for valuable fruit quality traits in Cucumis melo. Israel J. Plant Sci. 2006, 54, 233–242.

44. Beaulieu, J.C.; Grimm, C.C. Identification of volatile compounds in cantaloupe at various developmental stages using solid phase microextraction. J. Agric. Food Chem. 2001, 49, 1345–1352.

45. Jordán, M.J.; Shaw, P.E.; Goodner, K.L. Volatile components in aqueous essence and fresh fruit ofCucumis melo cv. Athena (muskmelon) by GC-MS and GC-O. J. Agric. Food Chem. 2001, 49, 5929–5933.

46. Bood, K.G.; Zabetakis, I. The biosynthesis of strawberry flavor (II): Biosynthetic and molecular biology studies. J. Food Sci. 2002, 67, 2–8.

47. Jetti, R.R.; Yang, E.; Kurnianta, A.; Finn, C.; Qian, M.C. Quantification of selected aroma-active compounds in strawberries by headspace solid-phase microextraction gas chromatography and correlation with sensory descriptive analysis. J. Food Sci. 2007, 72, S487–S496.

48. Schwab, W.; Schaart, J.G.; Rosati, C. molecular biology research in Fragaria. In Genetics and Genomics of Rosaceae; Folta, K.M., Gardiner, S.E., Eds.; Springer: New York, NY, USA, 2009.

49. Jayanty, S.; Song, J.; Rubinstein, N.M.; Chong, A.; Beaudry, R.M. Temporal relationship between ester biosynthesis and ripening events in bananas. J. Am. Soc. Hort. Sci. 2002, 127, 998–1005.

50. Wendakoon, S.K.; Ueda, Y.; Imahori, Y.; Ishimaru, M. Effect of short-term anaerobic conditions on the production of volatiles, activity of alcohol acetyltransferase and other quality traits of ripened bananas. J. Sci. Food Agric. 2006, 86, 1475–1480.

51. Boudhrioua, N.; Giampaoli, P.; Bonazzi, C. Changes in aromatic components of banana during ripening and air-drying. Lebensm Wiss Technol. 2003, 36, 633–642.

52. Aurore, G.; Ginies, C.; Ganou-Parfait, B.; Renard, C.M.G.C.; Fahrasmane, L. Comparative study of free and glycoconjugated volatile compounds of three banana cultivars from French West Indies: Cavendish, Frayssinette and Plantain. Food Chem. 2011, 129, 28–34.

53. Gonza'lez-Mas, M.C.; Rambla, J.L.; Alamar, M.C.; Antonio, A.G. Comparative analysis of the volatile fraction of fruit juice from differentCitrus species. PLoS One 2011, 6, e22016.

54. Kelebek, H.; Selli, S. Determination of volatile, phenolic, organic acid and sugar components in a Turkish cv. Dortyol (Citrus sinensis L. Osbeck) orange juice. J. Sci. Food Agric. 2011, 91, 1855–1862.

55. Tomiyama, K.; Aoki, H.; Oikawa, T.; Sakurai, K.; Kashara, Y.; Kawakami, Y. Characteristic volatile components of Japanese sour citrus fruits: Yuzu, Sudachi and Kabosu. Flavour Fragr. J. 2012, 27, 341–355.

56. Phi, N.T.L.; Nishiyama, C.; Choi, H.S.; Sawamura, M. Evaluation of characteristic aroma compounds of Citrus natsudaidai Hayata (Natsudaidai) cold-pressed peel oil. Biosci. Biotechnol. Biochem. 2006, 70, 1832–1838.

57. Shibamoto, T.; Tang, C.S. "Minor" tropical fruit mango, papaya, passion fruit, and guava. In Food Flavours: Part C: The Flavour of Fruit; Morton, I.D., MacLeod, A.J., Eds.; Elsevier: Amsterdam, The Netherlands, 1990; pp. 221–234.

58. Pino, J.A.; Mesa, J. Contribution of volatile compounds to mango (Mangifera indica L.) aroma. Flav. Frag. J. 2006, 21, 207–213.

59. Quijano, C.E.; Salamanca, G.; Jorge, A.; Pino, J.A. Aroma volatile consttuents of Colombian varieties of mango (Mangifera indica L.). Flav. Frag. J. 2007, 22, 401–406.

60. Narain, N.; Bora, P.S.; Narain, R.; Shaw, P.E. Mango. In Tropical and Subtropical Fruits; Shaw, P.E., Chan, H.T., Nagy, S., Eds.; AgScience: Auburndale, FL, USA, 1997; pp. 1–77.

61. Olle, D.; Baumes, R.L.; Bayonove, C.L.; Lozano, Y.F.; Sznaper, C.; Brillouet, J.M. Comparison of free and glycosidically linked volatile components from polyembryonic and monoembryonic mango (Mangifera indica L.) cultivars. J. Agric. Food Chem. 1998, 46, 1094–1100.

62. Aubert, C.; Milhet, C. Distribution of the volatile compounds in the different parts of a white-fleshed peach (Prunus persica L. Batsch). Food Chem. 2007, 102, 375–384.

63. Wang, Y.J.; Yang, C.X.; Li, S.H.; Yang, L.; Wang, Y.N.; Zhao, J.B.; Jiang, Q. Volatile characteristics of 50 peaches and nectarines evaluated by HP-SPME with GC-MS. Food Chem. 2009, 116, 356–364.

64. Aubert, C.; Gunata, Z.; Ambid, C.; Baumes, R. Changes in physicochemical characteristics and volatile constituents of yellow- and white-fleshed nectarines during maturation and artificial ripening. J. Agric. Food Chem. 2003, 51, 3083–3091.

65. Eduardo, I.; Chietera, G.; Bassi, D.; Rossini, L.; Vecchietti, A. Identification of key odor volatile compounds in the essential oil of nine peach accessions. J. Sci. Food Agric. 2010, 90, 1146–1154.

66. Zhang, B.; Shen, J.Y.; Wei, W.W.; Xi, W.P.; Xu, C.J.; Ferguson, I.; Chen, K.S. Expression of genes associated with aroma formation derived from the fatty acid pathway during peach fruit ripening. J. Agric. Food Chem. 2010, 58, 6157–6165.

67. Zhang, B.; Xi, W.P.; Wei, W.W.; Shen, J.Y.; Ferguson, I.; Chen, K.S. Changes in aroma-related volatiles and gene expression during low temperature storage and subsequent shelf-life of peach fruit.Postharvest Biol. Technol. 2011, 60, 7–16.

68. Nijssen, L.M.; Visscher, C.A.; Maarse, H.; Willemsens, L.C.; Boelens, M.H. Volatile Compounds in Foods Qualitative and Quantitative Data; TNO Nutrition and Food Research Institute: Zeist, The Netherlands, on-line version 9.2, 2007.

69. González-Agüero, M.; Troncoso, S.; Gudenschwager, O.; Campos-Vargas, R.; Moya-León, M.A.; Defilippi, B.G. Differential expression levels of aroma-related genes during ripening of apricot (Prunus armeniaca L.). Plant Physiol. Biochem. 2009, 47, 435–440.

70. Aubert, C.; Chanforan, C. Postharvest changes in physicochemical properties and volatile constituents of apricot (Prunus armeniaca L.). Characterization of 28 cultivars. J. Agric. Food Chem. 2007, 55, 3074–3082.

71. Dieguez, S.C.; Lois, L.C.; Gomez, E.F.; De Ia Pena, M.L.G. Aromatic composition of the Vitis viniferagrape Albariño. Lebensmittel-Wissenschaft und-Technologie 2003, 36, 585–590.

72. Rosilllo, L.; Salinas, M.R.; Garijo, J.; Alonso, G.L. Study of volatiles in grapes by dynamic headspace analysis Application to the differentiation of some Vitis vinifera varieties. J. Chromatogr. A 1999, 847, 155–159.

73. Bellincontro, A.; Nicoletti, I; Valentini, M.; Tomas, A.; de Santis, D.; Corradini, D.; Mencarelli, F. Integration of nondestructive techniques with destructive analyses to study postharvest water stress of winegrapes. Am. J. Enol. Vitic. 2009, 60, 57–65.

74. Fenoll, J.; Manso, A.; Hellin, P.; Ruiz, L.; Flores, P. Changes in the aromatic composition of the Vitis vinifera grape Muscat Hamburg during ripening. Food Chem. 2009, 114, 420–428.

75. Kalua, C.M.; Boss, P.K. Comparison of major volatile compounds from Riesling and Cabernet Sauvignon grapes (Vitis vinifera L.) from fruitset to harvest. Austra. J. Grape Wine Res. 2010, 16, 337–348.

76. Klesk, K.; Qian, M.; Martin, R. Aroma extract dilution analysis of cv. Meeker (Rubus idaeus L.) red raspberries from Oregon and Washington. J. Agric. Food Chem. 2004, 52, 5155–5161.

77. Malowicki, S.M.; Martin, R.; Qian, M.C. Volatile composition in raspberry cultivars grown in the Pacific Northwest determined

by stir bar sorptive extraction-gas chromatography-mass spectrometry.J. Agric. Food Chem. 2008, 56, 4128–4133.

78. Turemis, N.; Kafkas, E.; Kafkas, S.; Kurkcuoglu, M.; Baser, K.H.C. Determination of aroma compounds in blackberry by GC/MS analysis. Chem. Nat. Comp. 2003, 39, 174–176.

79. Du, X.F.; Finn, C.E.; Qian, M.C. Volatile composition and odour-activity value of thornless "Black Diamond" and "Marion" blackberries. Food Chem. 2010, 119, 1127–1134.

80. Du, X.F.; Plotto, A.; Song, M.; Olmstead, J.; Rouseff, R. Blueberry volatile composition of four southern highbush cultivars and effect of growing location and harvest date. J. Agric. Food Chem. 2011, 59, 8347–8357.

81. Tokitomo, Y.; Steinhaus, M.; Buttner, A.; Schieberle, P. Odor-active constituents in fresh pineapple (Ananas comosus [L.] Merr.) by quantitative and sensory evaluation. Biosci. Biotechnol. Biochem. 2005,69, 1323–1330.

82. Zhang, X.M.; Du, L.Q.; Sun, G.M.; Wei, C.B.; Liu, S.H.; Xie, J.H. Changes of aroma components in Yellow Mauritius pineapple during fruit development (In Chinese). J. Fruit Sci. 2009, 26, 245–249.

83. Liang, Y.Z.; Guang, M.S.; Yu, G.L.; Ling, L.L.; Wen, X.Y.; Wei, F.Z.; Chang, B.W. Aroma volatile compounds from two fresh pineapple varieties in China. Int. J. Mol. Sci. 2012, 13, 7383–7392.

84. Chang, B.W.; Sheng, H.L.; Yu, G.L.; Ling, L.L.; Wen, X.Y.; Guang, M.S. Characteristic aroma compounds from different pineapple parts. Molecules 2011, 16, 5104–5112.

85. Friel, E.N.; Wang, M.; Taylor, A.J.; MacRae, E.A. In vitro and in vivo release of aroma compounds from yellow-fleshed kiwifruit. J. Agric. Food Chem. 2007, 55, 6664–6673.

86. Gunther, C.S.; Matich, A.J.; Marsh, K.B.; Nicolau, L. (Methylsulfanyl) alkanoate ester biosynthesis in Actinidia

chinensis kiwifruit and changes during cold storage. Phytochemistry 2010, 71, 742–750.

87. Yang, C.; Wang, Y.; Liang, Z.; Fan, P.; Wu, B.; Yang, L.; Wang, Y.; Li, S. Volatiles of grape berries evaluated at the germplasm level by headspace-SPME with GC-MS. Food Chem. 2009, 114, 1106–1114.

88. Wang, H.B.; Chen, X.S.; Xin, P.G.; Feng, T.; Shi, J.; Ci, Z.J. GC-MS analysis of volatile components in several early apple cultivars (in Chinese). J. Fruit Sci. 2007, 24, 11–15.

89. Aharoni, A.; Giri, A.P.; Verstappen, F.W.A.; Bertea, C.M.; Sevenier, R.; Sun, Z.; Jongsma, M.A.; Schwab, W.; Bouwmeester, H.J. Gain and loss of fruit flavor compounds produced by wild and cultivated strawberry species. Plant Cell 2004, 16, 3110–3131.

90. Hampel, D.; Mosandl, A.; Wust, M. Biosynthesis of mono- and sesquiterpenes in strawberry fruits and foliage: 2H labeling studies. J. Agric. Food Chem. 2006, 54, 1473–1478.

91. Baldwin, E.A.; Scott, J.W.; Shewmaker, C.K.; Shuch, W. Flavor trivia and tomato aroma: biochemistry and possible mechanisms for control of important aroma components. HortScience 2000, 35, 1013–1022.

92. Shan, W.Y.; Zhao, C.; Fan, J.G.; Cong, H.; Liang, S.; Yu, X.Y. Antisense suppression of alcohol acetyltransferase gene in ripening melon fruit alters volatile composition. Sci. Hortic. 2012, 139, 96–101.

93. Lester, G. Consumer preference quality attributes of melon fruit. Acta Hortic. 2006, 712, 175–181.

94. Kader, A.A. A perspective on postharvest horticulture (1978–2003). Hort. Sci. 2004, 38, 1004–1008.

95. Fellman, J.K.; Rudell, D.R.; Mattinson, D.S.; Mattheis, J.P. Relationship of harvest maturity to flavor regeneration after CA storage of “Delicious” apples. Postharvest Biol. Technol. 2003, 27, 39–51.

96. Menager, I.; Jost, M.; Aubert, C. Changes in physicochemical characteristics and volatile constituents of strawberry (Cv. Cigaline) during maturation. J. Agric. Food Chem. 2004, 52, 1248–1254.

97. Beaulieu, J.C. Effect of cutting and storage on acetate and nonacetate esters in convenient, ready-to-eat fresh-cut melons and apples. Hort. Sci. 2006, 41, 65–73.

98. El-Mageed, M.A.A. Development of volatile compounds of avocado and casimiroa during fruit maturation. Arab Univ. J. Agric. Sci. 2007, 15, 89–99.

99. Pereira, M.E.C. Changes in volatile compounds during ripening of West Indian-type "Simmonds" avocado treated with ethylene and aqueous 1-methylcyclopropene. In American Society for Horticultural Science Annual Meeting; American Society for Horticultural Science: Palm Desert, CA, USA, 2010; p. S93.

100. Obenland, D.; Collin, S.; Sievert, J.; Negm, F.; Arpaia, M.L. Influence of maturity and ripening on aroma volatiles and flavor in "Hass" avocado. Postharvest Biol. Technol. 2012, 71, 41–50.

101. Fuggate, P.; Wongs-Aree, C.; Noichinda, S.; Kanlayanarat, S. Quality and volatile attributes of attached and detached "Pluk Mai Lie" papaya during fruit ripening. Sci. Hortic. 2010, 126, 120–129.

102. Yang, C.X.; Wang, Y.J.; Wu, B.H.; Fang, J.B.; Li, S.H. Volatile compounds evolution of three table grapes with different flavor during and after maturation. Food Chem. 2011, 128, 823–830.

103. Peyrot des Gachons, C.; Leeuwen, C.V.; Tominaga, T.; Soyer, J.P.; Gaudillere, J.P.; Denis, D. Influence of water and nitrogen deficit on fruit ripening and aroma potential of Vitis vinifera L cv Sauvignon blanc in field conditions. J. Sci. Food Agric. 2005, 85, 73–85.

104. Salas, N.A.; Molina-Corral, F.J.; Gonzalez-Aguilar, G.A.; Otero, A.; Sepulveda, D.R.; Olivas, G.I. Volatile production by "Golden

Delicious" apples is affected by preharvest application of aminoethoxyvinylglycine. Sci. Hortic. 2011, 130, 436–444.

105. Tietel, Z.; Lewinsohn, E.; Fallik, E.; Porat, R. Importance of storage temperatures in maintaining flavor and quality of mandarins. Postharvest Biol. Technol. 2012, 64, 175–182.

106. Sanchez, F.D.; Zaldivar, C.P.; Cabrera, F.R.; Valadez, M.P.; Alejandre, X.A.; Fernandez, F.J.; Buendiac, H.B.E.; Flores, L.J.P. Effect of refrigerated storage on aroma and alcohol dehydrogenase activity in tomato fruit. Postharvest Biol. Technol. 2009, 54, 93–100.

107. Beaulieu, J.C. Volatile changes in cantaloupe during growth, maturation, and in stored fresh-cuts prepared from fruit harvested at various maturities. J. Am. Soc. Hort. Sci. 2006, 131, 127–139.

108. Bai, J.; Baldwin, E.A.; Imahori, Y.; kostenyuk, I.; Burns, J.; Brecht, J.K. Chilling and heating may regulate C6 volatile aroma production by different mechanisms in tomato (Solanum lycopersicum) fruit.Postharvest Biol. Technol. 2011, 60, 111–120.

109. Mattheis, J.P.; Fan, X.T.; Argenta, L.C. Interactive responses of gala apple fruit volatile production to controlled atmosphere storage and chemical inhibition of ethylene action. J. Agric. Food Chem. 2005, 53, 4510–4516.

110. Ortiz, A.; Echeverria, G.; Lopez, M.L.; Graell, J.; Lara, I. Overall quality of "Rich Lady" peach fruit after air- or CA storage. The importance of volatile emission. LWT Food Sci. Technol. 2009, 42, 1520–1529.

111. Harb, J.; Bisharat, R.; Streif, J. Changes in volatile constituents of blackcurrants (Ribes nigrum L. cv. "Titania") following controlled atmosphere storage. Postharvest Biol. Technol. 2008, 47, 271–279.

112. Cohen, E.; Shalom, Y.; Rosenberger, I. Postharvest ethanol buildup and off-flavor in Murcott tangerine fruits. J. Am. Soc. Hort. Sci. 1990, 115, 775–778.

113. Dang, K.T.H.; Singh, Z.; Swinny, E.E. Edible Coatings Influence Fruit Ripening, Quality, and Aroma Biosynthesis in Mango Fruit. Agric. Food Chem. 2008, 56, 1361–1370.

114. Saftner, R.A.; Conway, W.S.; Sam, C.E. Postharvest calcium infiltration alone and combined with surface coating treatments influence volatile levels, respiration, ethylene production, and internal atmospheres of "Golden Delicious" apples. J. Am. Soc. Hort. Sci. 1999, 124, 553–558.

115. Baldwin, E.; Burns, J.K.; Kazokas, W; Brecht, J.K.; Hagenmaier, R.D.; Bender, R.J.; Pesis, E. Effect of two edible coatings with different permeability characteristics on mango (Mangifera indica L.) ripening during storage. Postharv. Biol. Technol. 1999, 17, 215–226.

116. McDonald, R.E.; McCollum, T.G.; Baldwin, E.A. Prestorage heat treatments influence free sterols and flavor volatiles of tomatoes stored at chilling temperature. J. Am. Soc. Hort. Sci. 1996, 12, 531–536.

117. Ortiz, A.; Echeverría, G.; Graell, J.; Lara, I. The emission of flavour-contributing volatile esters by "Golden Reinders" apples is improved after mid-term storage by postharvest calcium treatment.Postharvest Biol. Technol. 2010, 57, 114–123.

118. Lalel, H.J.D.; Singh, Z.; Tan, S.C. The role of methyl jasmonate in mango ripening and biosynthesis of aroma volatile compounds. J. Hortic. Sci. Biotechnol. 2003, 78, 470–484.

119. Kondo, S.; Mattheis, J. Aroma volatile biosynthesis in apples at harvest or after harvest affected by jasmonates. Acta Hortic. 2006, 712, 381–388.

120. De la Peña Moreno, F.; Blanch, G.P.; Ruiz del Castillo, M.L. Effect of (−)- and (+)-methyl jasmonate on the formation of aroma-

active esters in strawberry fruit. Eur. Food Res. Technol. 2010, 231, 829–834.

121. De la Peña Moreno, F.; Blanch, G.P.; Flores, G.; Ruiz del Castillo, M.L. Impact of postharvest methyl jasmonate treatment on the volatile composition and flavonol content of strawberries. J. Sci. Food Agric. 2010, 90, 989–994.

122. De la Peña Moreno, F.; Monagas, M.; Blanch, G.P.; Bartolomé, B.; Ruiz del Castillo, M.L. Enhancement of phenolic and aroma compounds in strawberry fruit through methyl jasmonate vapour treatment.Eur. Food Res. Technol. 2010, 230, 989–999.

123. Blanch, G.P.; Flores, G.; Castillo, M.L.R. Influence of methyl jasmonate in conjunction with ethanol on the formation of volatile compounds in berries belonging to the Rosaceae. Postharvest Biol. Technol.2011, 62, 168–178.

124. Dudareva, N.; Pichersky, E.; Gershenzon, J. Biochemistry of plant volatiles. Plant Physiol. 2004, 135, 1893–1902.

125. Gang, D.R. Evolution of flavors and scents. Annu. Rev. Plant Biol. 2005, 56, 301–325.

126. Song, J.; Bangerth, F. Fatty acids as precursors for aroma volatile biosynthesis in pre-climacteric and climacteric apple fruit. Postharvest Biol. Technol. 2003, 30, 113–121.

127. Schwab, W.; Schreier, P. Enzymic formation of flavor volatiles from lipids. In Lipid Biotechnology; Kuo, T.M., Gardner, H.W., Eds.; Marcel Dekker: New York, NY, USA, 2002; pp. 293–318.

128. Schreier, P. Chromatographic Studies of Biogenesis of Plant Volatiles; Alfred Hüthig Verlag GmbH: Heidelberg, Germany, 1984.

129. Chan, H.W.S. Autoxidation of Unsaturated Lipids; Academic Press: London, UK, 1987.

130. Baker, A.; Graham, I.A.; Hodsworth, M.; Smith, S.M. Chewing the fat: β-Oxidation in signalling and development. Trends Plant Sci. 2006, 11, 124–132.

131. Goepfert, S.; Poirier, Y. β-Oxidation in fatty acid degradation and beyond. Curr. Opin. Plant Biol. 2007,10, 245–251.

132. Bartley, I.M.; Stoker, P.G.; Martin, A.D.E.; Hatfield, S.G.S.; Knee, M. Synthesis of aroma compounds by apples supplied with alcohols and methyl esters of fatty acids. J. Sci. Food Agric. 1985, 36, 567–574.

133. Paillard, N.M.M. The flavour of apples, pears and quinces. In Food Flavours, Part C: The Flavour of Fruits; Morton, L.D., MacLeod, A.J., Eds.; Elsevier Science Publishing Company Inc.: Amsterdam, The Netherlands, 1990; pp. 1–41.

134. Tressl, R.; Albrecht, W. Biogenesis of aroma compounds through acyl pathways. In Biogeneration of Aromas; Parliament, T.H., Croteau, R., Eds.; ACS: Washington, DC, USA, 1986; pp. 114–133.

135. Perez, A.G.; Sanz, C. Formation of fruit flavor. In Fruit and Vegetable Flavour; Bruckner, B., Wyllie, S.G., Eds.; CRC Press: Boca Raton, FL, USA, 2008; pp. 71–102.

136. Matsui, K. Green leaf volatiles: Hydroperoxide lyase pathway of oxylipin metabolism. Curr. Opin. Plant Biol. 2006, 9, 274–280.

137. Stumpe, M.; Feussner, I. Formation of oxylipins by CYP74 enzymes. Phytochem. Rev. 2006, 5, 347–357.

138. Lea, A.G.H. Apple juice. In Production and Packaging of Non-Carbonated Fruit Juices and Fruit Beverages, 3rd ed.; Ashurts, P.R., Ed.; Springer: Berlin, Germany, 1995; pp. 153–196.

139. De Pooter, H.L.; Montens, J.P.; Dirinck, J.; Willaert, G.A.; Schamp, N.M. Treatment of Golden Delicious apple with aldehydes and carboxylic acids: Effect on the headspace composition. J. Agric. Food Chem. 1983, 37, 813–818.

140. Feussner, I.; Wasternack, C. The lipoxygenase pathway. Annu. Rev. Plant Biol. 2002, 53, 275–297.

141. Song, M.S.; Kim, D.G.; Lee, S.H. Isolation and characterization of a jasmonic acid carboxyl methyltransferase gene from hot pepper (Capsicum annuum L.). J. Plant Biol. 2005, 48, 292–297.

142. Akacha, N.B.; Boubaker, O.; Gargouri, M. Production of hexenol in a two-enzyme system: Kinetic study and modelling. Biotechnol. Lett. 2005, 27, 1875–1878.

143. Baldwin, I.T.; Kessler, A.; Halitschke, R. Volatile signaling in plantplant-herbivore interactions: What is real. J. Curr. Opin. Plant Biol. 2002, 5, 351–354.

144. Beck, H.C.; Hansen, A.M.; Lauritsen, F.R. Metabolite production and kinetics of branched-chain aldehyde oxidation in Staphylococcus xylosus. Enzyme Microb. Technol. 2002, 31, 94–101.

145. Tavaria, F.K.; Dahl, S.; Carballo, F.J.; Malcata, F.X. Amino acid catabolism and generation of volatiles by lactic acid bacteria. J. Dairy Sci. 2002, 85, 2462–2470.

146. Perez, A.G.; Olias, R.; Luaces, P.; Sanz, C. Biosynthesis of Strawberry Aroma Compounds through Amino Acid Metabolism. J. Agric. Food Chem. 2002, 50, 4037–4042.

147. Rowan, D.D.; Lane, H.P.; Allen, J.M.; Fielder, S.; Hunt, M.B. Biosynthesis of 2-methylbutyl, 2-methyl-2-butenyl, and 2-methylbutanoate esters in red delicious and granny smith apples using deuterium-labeled substrates. J. Agric. Food Chem. 1996, 44, 3276–3285.

148. Wyllie, S.G.; Fellman, J.K. Formation of volatile branched chain esters in bananas (Musa sapientum L.).J. Agric. Food Chem. 2000, 48, 3493–3496.

149. Perez, A.G.; Rios, J.J.; Sanz, C.; Olias, J.M. Aroma components and free amino acids in strawberry variety Chandler during ripening. J. Agric.Food Chem. 1992, 40, 2232–2235.

150. Beekwilder, J.; Alvarez-Huerta, M.; Neef, E.; Verstappen, F.W.A.; Bouwmeester, H.J.; Aharoni, A. Substrate usage by recombinant alcohol acyltransferases from various fruit species. Plant Physiol. 2004,135, 1865–1878.

151. Newman, J.D.; Chappell, J. Isoprenoid biosynthesis in plants: Carbon partitioning within the cytoplasmic pathway. Crit. Rev. Biochem. Mol. Biol. 1999, 34, 95–106.

152. Lichtenthaler, H.K. The 1-deoxy-D-xylulose-5-phosphate pathway of isoprenoid biosynthesis in plants.Annu. Rev. Plant Physiol. Plant Mol. Biol. 1999, 50, 47–65.

153. Dudareva, N.; Andersson, S.; Orlova, I.; Gatto, N.; Reichelt, M.; Rhodes, D. The nonmevalonate pathway supports both monoterpene and sesquiterpene formation in snapdragon flowers. Proc. Natl. Acad. Sci. USA 2005, 102, 933–938.

154. Laule, O.; Furholz, A.; Chang, H.S.; Zhu, T.; Wang, X.; Heifetz, P.B. Crosstalk between cytosolic and plastidial pathways of isoprenoid biosynthesis in Arabidopsis thaliana. Proc. Natl. Acad. Sci. USA 2003,100, 6866–6871.

155. Miller, B.; Oschinski, C.; Zimmer, W. First isolation of an isoprene synthase gene from poplar and successful expression of the gene in Escherichia coli. Planta 2001, 213, 483–487.

156. Rodriguez-Concepcion, M.; Boronat, A. Elucidation of the methylerythritol phosphate pathway for isoprenoid biosynthesis in bacteria and plastids. A metabolic milestone achieved through genomics.Plant Physiol. 2002, 130, 1079–1089.

157. McGarvey, D.J.; Croteau, R. Terpenoid metabolism. Plant Cell 1995, 7, 1015–1026.

158. Ogura, K.; Koyama, T. Enzymatic aspects of isoprenoid chain elongation. Chem. Rev. 1998, 98, 1263–1276.

159. Koyama, T.; Ogura, K. Isopentenyl diphosphate isomerase and prenyltransferases. In Comprehensive Natural Products Chemistry: Isoprenoids Including Carotenoids and Steroids; Cane, D.E., Ed.; Pergamon Press, Oxford, UK, 1999; Volume 2, pp. 69–96.

160. Cane, D.E. Sesquiterpene biosynthesis: Cyclization mechanisms. In Comprehensive Natural Products Chemistry: Isoprenoids

Including Carotenoids and Steroids; Cane, D.E., Ed.; Pergamon Press: Oxford, UK, 1999; Volume 2, pp. 155–200.

161. Wise, M.L.; Croteau, R. Monoterpene biosynthesis. In Comprehensive Natural Products Chemistry: Isoprenoids Including Carotenoids and Steroids; Cane, D.E., Ed.; USA: Pergamon Press: Oxford, UK, 1999; Volume 2, pp. 97–153.

162. Martin, D.; Faldt, J.; Bohlmann, J. Functional characterization of nine Norway spruce TPS genes and evolution of gymnosperm terpene synthases of the TPS-d subfamily. Plant Physiol. 2004, 135, 1908–1927.

163. Lupien, S.; Karp, F.; Wildung, M.; Croteau, R. Regiospecific cytochrome P450 limonene hydroxylases from mint (Mentha) species: Cdna isolation, characterization, and functional expression of (−)-4S-limonene-3-hydroxylase and (−)-4S-limonene-6-hydroxylase. Arch. Biochem. Biophys. 1999, 368, 181–192.

164. Bouwmeester, H.J.; Gershenzon, J.; Konings, M.C.J.M.; Croteau, R. Biosynthesis of the monoterpenes limonene and carvone in the fruit of caraway—I. Demonstration of enzyme activities and their changes with development. Plant Physiol. 1998, 117, 901–912.

165. Bouwmeester, H.J.; Konings, M.C.J.M.; Gershenzon, J.; Karp., F.; Croteau, R. Cytochrome P-450 dependent (+)-limonene-6-hydroxylation in fruits of caraway (Carum carvi). Phytochemistry 1999, 50, 243–248.

166. Iijima, Y.; Wang, G.; Fridman, E.; Pichersky, E. Analysis of the enzymatic formation of citral in the glands of sweet basil. Arch. Biochem. Biophys. 2006, 448, 141–149.

167. Shalit, M.; Guterman, I.; Volpin, H.; Bar, E.; Tamari, T.; Menda, N.; Adam, Z.; Zamir, D.; Vainstein, A.; Weiss, D.; et al. Volatile ester formation in roses: Identification of an acetyl-CoA: Geraniol acetyltransferase in developing rose petals. Plant Physiol. 2003, 131, 1868–1876.

168. Bauer, K.; Garbe, D.; Surburg, H. Common Fragrance and Flavor Materials; Wiley-VCH Velagsgesellschaft mbH: Weinheim, Germany, 2001.

169. Luan, F.; Mosandl, A.; Munch, A.; Wust, M. Metabolism of geraniol in grape berry mesocarp of Vitis vinifera L. cv. Scheurebe: Demonstration of stereoselective reduction, E/Z-isomerization, oxidation and glycosylation. Phytochemistry 2005, 66, 295–303.

170. Degenhardt, J.; Gershenzon, J. Demonstration and characterization of (E)-nerolidol synthase from maize: A herbivore-inducible terpene synthase participating in (3E)-4,8-dimethyl-1,3,7-nonatriene biosynthesis. Planta 2000, 210, 815–822.

171. Rosati, C.; Diretto, G.; Giuliano, G. Biosynthesis and Engineering of Carotenoids and Apocarotenoids in Plants: State of the Art and Future Prospects. Biotech. Genet. Eng. Rev. 2009, 26, 151–174.

172. Auldridge, M.E.; McCarty, D.R.; Klee, H.J. Plant carotenoid cleavage oxygenases and their apocarotenoid products. Curr. Opin. Plant Biol. 2006, 9, 315–321.

173. Simkin, A.J.; Schwartz, S.H.; Auldridge, M.; Taylor, M.G.; Klee, H.J. The tomato carotenoid cleavage dioxygenase 1 genes contribute to the formation of the flavor volatiles beta-ionone, pseudoionone, and geranylacetone. Plant J. 2004, 40, 882–892.

174. Rubio, A.; Rambla, J.L.; Santaella, M.; Gomez, M.D.; Orzaez, D.; Granell, A.; Gomez-Gomez, L. Cytosolic and plastoglobule-targeted carotenoid dioxygenases from Crocus sativus are both involved in betai onone release. J. Biol. Chem. 2008, 283, 24816–24825.

175. Auldridge, M.E.; Block, A.; Vogel, J.T.; Dabney-Smith, C.; Mila, I.; Bouzayen, M.; Magallanes-Lundback, M.; DellaPenna, D.; McCarty, D.R.; Klee, H.J. Characterization of three members of the Arabidopsis carotenoid cleavage dioxygenase family

demonstrates the divergent roles of this multifunctional enzyme family. Plant J. 2006, 45, 982–993.

176. Simkin, A.J.; Underwood, B.A.; Auldridge, M.; Loucas, H.M.; Shibuya, K.; Schmelz, E.; Clark, D.G.; Klee, H.J. Circadian regulation of the PhCCD1 carotenoid cleavage dioxygenase controls emission of beta-ionone, a fragrance volatile of petunia flowers. Plant Physiol. 2004, 136, 3504–3514.

177. Booker, J.; Auldridge, M.; Wills, S.; McCarty, D.; Klee, H.J.; Leyser, O. MAX3/CCD7 is a carotenoid cleavage dioxygenase required for the synthesis of a novel plant signaling molecule. Curr. Biol. 2004,14, 1232–1238.

178. Schwartz, S.H.; Qin, X.; Loewen, M.C. The biochemical characterization of two carotenoid cleavage enzymes from Arabidopsis indicates that a carotenoid-derived compound inhibits lateral branching. J. Biol. Chem. 2004, 279, 46940–46945.

179. Bouvier, F.; Dogbo, O.; Camara, B. Biosynthesis of the food and cosmetic plant pigment bixin (annatto). Science 2003, 300, 2089–2091.

180. Bouvier, F.; Suire, C.; Mutterer, J.; Camara, B. Oxidative remodeling of chromoplast carotenoids: identification of the carotenoid dioxygenase CsCCD and CsZCD genes involved in Crocus secondary metabolite biogenesis. Plant Cell 2003, 15, 47–62.

3

EXTENSION OF SMALL-SCALE POSTHARVEST HORTICULTURE TECHNOLOGIES—A MODEL TRAINING AND SERVICES CENTER

Lisa Kitinoja[1] and Diane M. Barrett[2]

[1]World Food Logistics Organization, 1500 King Street, Alexandria, VA 22314, USA

[2]Department of Food Science and Technology, University of California, Davis, One Shields Ave, Davis, CA 95616, USA

ABSTRACT

A pilot Postharvest Training and Services Center (PTSC) was launched in October 2012 in Arusha, Tanzania as part of a United States Agency for International Development (USAID) funded project. The five key components of the PTSC are (1) training of postharvest trainers, (2) postharvest training and demonstrations for local small-scale clientele, (3) adaptive research, (4) postharvest services, and (5) retail sales of postharvest tools and supplies. During the years of 2011–2012, a one year e-learning program was provided to 36 young horticultural professionals from seven Sub-Saharan African countries.

These postharvest specialists went on to train more than 13,000 local farmers, extension workers, food processors, and marketers in their home countries in the year following completion of their course. Evaluators found that these specialists had trained an additional 9300 people by November 2014. When asked about adoption by their local trainees, 79% reported examples of their trainees using improved postharvest practices. From 2012–2013, the project supported 30 multi-day training programs, and the evaluation found that many of the improved practices being promoted were adopted by the trainees and led to increased earnings. Three PTSC components still require attention. Research activities initiated during the project are incomplete, and successful sales of postharvest goods and services will require commitment and improved partnering.

Keywords: postharvest technologies; e-learning; small-scale; impact evaluation; Sub-Saharan Africa

1. INTRODUCTION

A USAID-funded project in East Africa of a new model for extension of postharvest technologies was piloted during 2010–2014 under the Horticulture CRSP (Collaborative Research Support Program, now known as the Horticulture Innovation Lab, Davis, CA, USA) at the University of California (UC) at Davis. Project partners included the Department of Food Science and Technology at UC Davis, the International Programs office at the World Food Logistics Organization (WFLO, Alexandria, VA, USA), the University of Georgia, and the Asian Vegetable Research Development Center (AVRDC)—The World Vegetable Center (Arusha, Tanzania), and additional financial support and in-kind technical support was provided by postharvest specialists from The Postharvest Education Foundation (PEF), in which the authors are founding board members.

The pilot Postharvest Training and Services Center (PTSC) was launched in October 2012 on the campus of AVRDC—The World Vegetable Center in Arusha, Tanzania. The project operated as planned in Northern Tanzania (2012–2014), training young horticultural professionals from seven countries (Benin, Ethiopia, Ghana, Kenya, Rwanda, Tanzania, and Uganda) as "Postharvest

Specialists" and was followed six months later by USAID Save the Children/Technical and Operational Performance Support (TOPS)-funded evaluation studies (October 2014–May 2015). Simultaneously, during 2012–2015, three years of mentoring were being provided by PEF for more than a dozen Tanzanian extension workers and postharvest trainers who are currently working within the local community in and near Arusha to improve postharvest handling practices and reduce losses in horticultural crops.

1.1. Background Information

Less than 5% of funding for horticultural research and extension (R&E) has been allocated to postharvest issues over the past 25 years [1,2] as the historical focus has been on increasing production. In the 1990s, the focus in the horticultural research sector moved to the field of marketing and more recently it has focused upon value chain development. Internet database searches show that less than one in 2000 agricultural projects undertaken globally have focused on fresh produce handling and marketing, according to advanced searches undertaken in five major online databases during 2010 (AidData [3]; USAID Documents [4]; World Bank [5]; United Nations Food and Agricultural Organization (UN FAO) Information Network for Postharvest Operations (INPHo) [6]; Devex [7]). While thousands of development projects have been launched in developing countries between 1990 and the present time by dozens of donor agencies and governmental bodies, few have focused on agriculture (less than 6% according to the AiDA database [3]; 25% according to the World Bank [5]), very few have focused on horticulture (approximately 1% of the agricultural projects), and only one-third of this 1% of horticultural projects included any kind of postharvest component [8].

Of the 1.3 billion tons of food losses and waste reported by the UN FAO [9], an external analysis shows that 44% is made up of fruits and vegetable crops, and 20% is roots and tuber crops [10]. In terms of the percent of kilocalories (kcal) lost or wasted for each type of food commodity, roots and tubers experience the greatest amount of loss and waste (63% on a caloric basis) while the rate of kcal losses for fruits and vegetables is 42%. In comparison, about 25% of the kcal of cereals and of seafood produced are lost or wasted.

A UN FAO–commissioned report published in 2011 advocated for "strengthening the supply chain through training and support for farmers, making investments in infrastructure and transportation, as well as investments in an expansion of the food processing and packaging industries", which experts believe could help to reduce the amount of food loss and waste [9]. In most countries where high levels of fresh produce wastage is occurring, the local farmers, traders, small-scale processors, and marketers of fruits and vegetable crops have little or no access to postharvest training, technical information, guidance on use of new technologies, or local access to the tools and supplies needed to utilize new technical knowledge or improved postharvest handling practices.

In developing countries, food losses and food waste occur mainly at earlier stages of the value chain and can be traced back to financial, managerial, and technical constraints in harvesting techniques and postharvest handling as well as a lack of storage and cooling facilities. While postharvest losses for fruits and vegetable crops are reported to be 44% of the total global production by weight [10], these losses have been measured in Sub-Saharan Africa (SSA) for many crops to be even higher levels of 40% to 80% [11].

1.2. Description of the Model

The five key components of the postharvest extension model known as the PTSC are (1) training of postharvest trainers, (2) postharvest training demonstrations for local clientele, (3) adaptive research, (4) postharvest services, and (5) retail sales of tools and postharvest supplies. The concept of the PTSC was developed during 2007–2008 by the lead author while she was serving as an independent consultant for a Millennium Challenge Corporation (MCC) project in Cape Verde. Several PTSCs were constructed in Cape Verde during 2008–2010, but the designs were "enhanced" by the local authorities to include beautiful architectural features, millipede inspection stations, and/or large-scale cold storage facilities, so by the time all the modifications were made, the costs were very high and the design was not promoted beyond the MCC project for the Cape Verde islands. The original concept and model were further developed as part of a research project for the Bill and Melinda Gates Foundation which the lead author developed and led for WFLO during 2009–2010 on the investigation of appropriate postharvest technologies that could be disseminated in a sustainable manner in SSA and South Asia.

Postharvest advocacy needs in developing countries are many and include enhanced funding for R&E, updating laws governing markets, access to micro-credit, reduced interest rates, support for associations, extension programs for women, hiring and training more women as horticultural extension agents, and access to high quality planting materials/seeds, simple postharvest tools, supplies (especially improved packages), equipment, and market information [8]. Training of trainers, building local postharvest extension capacity, and training local populations in the specialized skills needed for reducing food losses have been advocated and proposed for many years, but very few studies have been funded or implemented. This pilot project was the first attempt to bring all the needed components together in one location and provide potential users with a complete package of postharvest training, supplies, services, and support activities.

The PTSC was designed to serve as:

- a site for extension workers and local postharvest trainers to meet with growers and others working along the value chains to provide training to improve local capacity and knowledge on improved produce handling, harvesting, sorting/grading, packing, cooling, storage, food safety, processing, and marketing practices
- a training venue with permanent demonstrations for observing improved, cost-effective small-scale postharvest handling practices, facilities, and equipment
- a site where local private companies can demonstrate and explain the benefits of their goods and services related to improved postharvest handling, processing, or storage

- a retail shop with postharvest tools and supplies, packages, plastic crates, and other goods that can be purchased locally at reasonable prices
- a place where people can come to ask questions or get advice on how to use improved postharvest practices, learn about costs and benefits and marketing options
- a place where growers or traders can pay a small fee for services such as having their produce packed in improved containers, cooled and/or stored for a few days before marketing, leasing of a small insulated transport vehicle, using a solar dryer to produce dried fruits or vegetable snack products, *etc.*

2. THE EVALUATION PLAN AND OBJECTIVES

The final project report for the Horticulture CRSP project was submitted by WFLO in May 2014. The evaluation team reviewed the final report, and the objectives for an *ex post facto* evaluation of a completed postharvest extension project were set in cooperation with USAID/TOPS, and the data collection and analysis methods were planned during the proposal phase of the project.

2.1. Data Collection Methods

Data collection instruments (written surveys and structured interview schedules) were developed and field-tested in SSA by WFLO's local evaluation team, and data was collected over a six-month period via email surveys, phone, and face-to-face interviews to characterize the implementation, outcomes, and impacts of the five project components of the PTSC pilot project. Written surveys and questionnaires were developed using traditional evaluation methods, in collaboration with stakeholders, the local evaluation team, and clientele in order to ensure the questions were suitable and easy to understand, and that responses would be reliable [12]. Table 1 provides a summary of the data collection plan, target groups, sources of data, and response rate.

2.2. Objectives of the Ex Post Facto Evaluation

The evaluation project had three major objectives.

- Objective 1: To determine the major capacity building outcomes and impact of the Postharvest Training and Services Center (PTSC) and Training the Trainers (ToT) program
- Objective 2: To identify best practices in the management of the PTSC and its extension services
- Objective 3: To identify problems, concerns, and obstacles to making the PTSC a sustainable and replicable model

Data was collected using a variety of methods (surveys, face-to-face or phone interviews, site visits) as shown in Table 1, where x indicates which method(s) were used for which target groups.

Table 1. Data collection plan and implementation.

Target Groups	Size of Target Group	Email Surveys	Face-to-Face Interviews	Phone Interviews	Site Visits for Observations	Response Rate
Training of trainers (ToT) participants	36 in total	x				92%
Local trainees (farmers and food processors)	50 people, random cluster sample selected from a population of 500 trainees		x		x	100%
Postharvest trainers	14 in total	x		x		100%
PTSC administrators and managers	7 in total		x		x	100%

3. RESULTS AND DISCUSSION

Data analyses undertaken for the *ex post facto* evaluation project included simple counts and percentages for quantitative data and descriptive information (categories, lists, case studies) for qualitative data. This communication draws on both the final project report for Hort CRSP and preliminary evaluation reports for USAID/TOPS, none of which have been published. Project leaders at WFLO plan to formally present the results of the TOPS-funded postharvest extension project evaluation once the project has been completed.

3.1. Training of Postharvest Trainers

One of the major objectives of the Hort CRSP–funded project was to educate 30 trainers as postharvest specialists and provide them with the knowledge, tools, and motivation to train 5000 local clientele in their own home locations by the end of the project. The project outcomes exceeded these targets, as 36 young horticultural professionals from seven countries in SSA were provided with 18 months of e-learning–based training during 2011–2012. All 36 completed the postharvest e-learning program and were given certificates as postharvest specialists. Topics included postharvest loss assessment, commodity systems assessment, technical information on small-scale practices, designing postharvest demonstrations (on harvesting, handling, sorting/grading, improved packages, cooling practices, storage, food processing methods, and more), planning extension programs, and designing their own PTSCs for their home countries.

From October 2012 through October 2013, without any project-provided funding but with access to mentoring and technical support, these 36 trainers planned and implemented postharvest horticultural training programs for more than 13,000 local farmers, extension workers, food processors, and marketers in their home countries. The TOPS evaluation reached 33 of these trainers, and their reports revealed that from November 2013 to November 2014, an additional 9300 people were trained by 28 of the trainers. The five persons who did not report providing any postharvest training for local populations were either actively pursuing graduate studies (in Germany, South Africa, and the USA) or had taken new jobs outside the agricultural sector (in Ghana and in Tanzania).

When asked about adoption of postharvest practices by their local trainees, 79% of the respondents reported on examples of trainees who are now using improved postharvest practices. Among those practices were improved harvesting, packing practices, and postharvest handling practices, Zero Energy Cool Chamber (ZECC) storage, solar drying, use of a cool room (via second-hand reefer container), postharvest loss assessment, and use of a postharvest tool kit and materials. However, solar drying and improved packaging practices were the most commonly used practices. Some farmers and traders of fresh horticultural crops started to dry and package their surplus produce or products which are not sold at the daily market with improved (vented) plastic bags, in order to avoid postharvest losses. These simple practices provide value in addition to the commodity and allow trainees to increase their incomes.

3.2. Postharvest Training and Demonstrations for Local Farmers in Tanzania

During 2012–2013, 14 instructors provided more than 30 postharvest training programs in Tanzania for local farmers, food processors, and marketers. Nine of the 14 postharvest instructors were women (from the USA, Lebanon, New Zealand, and Tanzania), and five were men (from Uganda, the USA, Zimbabwe, and Tanzania). Most of the instructors were independent consultants working with local organizations in Tanzania. Four of the instructors (two men and two women) were recent graduate "Postharvest Specialists" of the Hort CRSP ToT program. All 14 instructors were evaluated during the TOPS project. They were each contacted via email by one of WFLO's local consultants, and she then followed up via phone calls to probe for details and any missing information and to ensure that each written survey was fully completed.

A few of the training programs were open to the public ("open house days" where local people were welcome to attend), but most were designed for a specific audience, based on requests or locally assessed training needs. Most of the training programs were offered over a period of several days, and all of the programs covered three or more postharvest topics. The sites for training included the PTSC at AVRDC, the Ministry of Agriculture, Food Security and Cooperatives (MAFC) Njiro training venue, and the OIKOS Mkuru Training Camp.

Individuals or groups were invited by telephone or a group would make inquiries at AVRDC or the Selian Agricultural Research Institute (SARI) in Arusha about having a postharvest training program developed for them. According to these 14 instructors, the topics that women asked about most often were the use of shade, gentle handling, home storage, food processing, and marketing. The topics asked about most often by men were cooling, cold storage, transport, Good Agricultural Practices (GAPs), food safety, food processing, and marketing.

Two WFLO consultants, based locally in Arusha, conducted face-to-face interviews with 50 participants from the training programs. This sample was comprised of five persons who had been randomly selected from 10 randomly selected participant groups via cluster sampling. The survey results included responses from 42 women and eight men who had participated in PTSC training programs, with an average age of 46 years. The topics of the training programs were improved postharvest handling of fresh produce, ZECC storage, food processing methods (jam making, solar drying), cooling/cold storage, or marketing of horticultural crops. Most of the respondents identified themselves as farmers or food processors, and many were also marketers of their own fresh produce or processed products. Two of the respondents were also working as extension workers.

A wide range of demonstrations were mounted during the Hort CRSP project and have continued since the project ended. Specifications for demonstrations were based on published research and review articles [11,13,14,15].

These include:

- Use of shade
- Hand-washing/hygiene practices
- Improved containers (plastic crates, half-size wooden crates)
- Zero energy cool chamber (for storage of fresh fruits and vegetables)

- Solar drying
- Cold room equipped with CoolBot

Several new demonstrations have been added during 2014–2015 by the postharvest staff at AVRDC. These include:

- A simple hydro-cooler with recirculation system
- Improved wooden crates (smaller, smooth on the inside)
- Liners for use in traditional containers
- Wakati (a high relative humidity % storage container; under study with Arne Pauwels of Belgium)

Because the trainings took place at the PTSC sites where these demonstrations have been established, the 50 respondents reported having seen many of the following demonstrations (see Table 2 for a count of the number indicating they had seen each of the demonstrations). In addition, when they were asked to rate the usefulness (as most or least useful), most of the respondents did not select one demonstration as was anticipated, but each rated many of the demonstrations they had seen as "most useful".

Only three of the 50 respondents reported that they had not used one or more of the practices that they had first seen in a postharvest demonstration. Many reported using maturity indices (25), shade (27), sorting/grading (40), and/or improved containers (34) to help reduce fresh produce wastage.

The demonstrations designated as "least useful" were those with a high perceived cost (a cold room equipped with the CoolBot™, at a cost of approximately $2000 for specialized equipment and materials for self-construction, had not been adopted by any of the respondents, and the equipment required for jam-making was deemed too expensive by one respondent). Of the practices rated as "most useful", the least-cited was hand-washing and hygiene, which is already being practiced by those doing food processing.

All of the 50 persons interviewed for the evaluation indicated that they had adopted new postharvest handling practices or technologies for reducing losses and increasing the value of their crops after receiving training via the PTSC, and 42 people were able to provide details on local costs and benefits.

Table 2. Training participant ratings of postharvest demonstrations.

Demonstrations	No. that have seen it	Rated the demo as most useful	Rated the demo as least useful	No. that have been using the new or improved practice
Shade	44	31		27
Gentle handling	33	24		17
Maturity indices	33	24		25
Improved containers	45	32		34
Sorting/grading	47	36		40
Hand-washing/hygiene	45	10		35
ZECC	45	33		20
CoolBot™	37	3	24	0
Solar drying	44	34		29
Jam making	43	32	1	17

3.3. Adaptive Research

Adaptive research was planned on pest control, low-cost cool chambers, improved solar dryers, cool transport in insulated containers, food safety and/or other topics, but the research studies were not successfully carried out by AVRDC. Several meetings were held on the planning process during 2012 and 2013, and visits were made to Africa by the Principal Investigators and the Hort Innovation Lab management entity team in order to kick-start the process. In June 2013, a no-cost extension was granted to AVRDC in order to allow them more time to carry out some of these research studies.

In February 2014, AVRDC hired independent consultants to complete as much of the work as possible. Research studies were conducted on improved traditional containers (liners in wooden crates, use of plastic crates), consumer packages (recyclable clamshells and very thin plastic produce bags), insulated pallet covers for use during shipping (in order to measure effects on temperature change), and solar dryer modifications for enhancing drying during overcast weather (adding a black plastic wrap around the legs at the bottom of the dryer). AVRDC managed the work and provided the needed funds. A few of the studies have since been completed, and the results of a study on the use of low-cost insulated pallet covers and consumer packages for reducing losses of amaranth has been submitted to a major journal.

3.4. Postharvest Services at the PTSC in Arusha

Advisory services for those who are interested in adopting new practices and technologies were one of the only services being provided by the PTSC since the close of the project in 2014. The TOPS evaluation project revealed that since the end of the Hort CRSP pilot project, several local communities and training groups near Arusha have been receiving advice on constructing ZECCs for storing fresh foods on their farms or for food service. The ToT participants and AVRDC postharvest staff in Tanzania are actively making advisory field visits, meeting interested groups, and providing consulting for local clientele. Several local non-governmental organizations (NGOs) (Istituto OIKOS, The Mesula Project, ECHO-Impact Center, Arusha, Tanzania) have requested assistance and received postharvest training and advice on setting up their own demonstrations and training programs. None of the other planned "fee for services" activities (sorting/grading, packing, pre-cooling, cold storage, *etc.*) were being provided by the PTSC for local clientele.

The model included a variety of these postharvest services which were intended to serve as a ready source of income, generating funds to pay for utilities, management, and maintenance staff for the facilities. Without these sources of revenue, the PTSC will remain dependent upon the host organization for funding its management and training programs.

3.5. Retail Sales of Tools, Packages, and Postharvest Supplies

The evaluation revealed that the retail shop for the PTSC was never fully implemented, and much of the original inventory of tools and supplies provided by the Hort CRSP project is still on the shelves. The PTSC shop mainly functions for one week per year during the August "Nane-Nane" agricultural show at Njiro, Tanzania, when the shop is restocked with the most desired products and opened for business during a period of eight days so people can visit to make their purchases.

Both customers and the administrators of the PTSC project considered the retail shop to be one of the services provided for the local population, and expected the prices of goods and supplies to be kept lower than those at any competing vendor. The PTSC shop managers, on the other hand, did not have any wholesale buying expertise, and did not have the budget to be able to buy items in large enough quantities to make bulk purchases for the shop at lower wholesale prices.

When the 50 training program participants were queried regarding where they would obtain tools, supplies, and training if the PTSC did not exist, more than 50% said "nowhere" or said they "didn't know" or "were not sure" (Table 3).

Table 3. Respondent (n = 50) answers to queries on where to obtain postharvest tools, supplies, and training.

If the PTSC did not exist:	Where would you go for postharvest training?	Where would you go for postharvest demos/advice?	Where would you buy postharvest goods and services?
Nowhere	20	8	0
I don't know	7	15	3
Not sure	0	8	2
Other sources/charity organizations	21	16	16

The other sources listed by respondents included several local NGOs, churches, and international charities, but these sources were thought to be unreliable since they seldom offered training and did not always have the needed supplies. For example, OIKOS was included as a possible source for glass jars, but the evaluators learned that OIKOS buys its jars from the PTSC. The evaluation results make it clear that the PTSC shop has an important role to play in local postharvest loss reduction and small business development in Northern Tanzania, if it could be set up, stocked, operated, and marketed with a more business-like approach.

4. TIMELINES AND BUDGETS

The PTSC set-up process began in early 2012 on the AVRDC campus in Arusha, Tanzania. AVRDC hired a postharvest specialist in June 2012, and assigned one of their campus staff as the PTSC manager in July. The PTSC designs, demonstration protocols, procurement lists, and equipment/supplies specifications were all provided by the project leaders. Renovations and procurements went over budget, but AVRDC paid the additional costs since they intended to use the PTSC for their own projects and programs as well as for Hort CRSP–funded project activities.

The total budget for the Hort CRSP project was $500,000, with approximately $100,000 for implementing the ToT program, and approximately $100,000 used for the PTSC renovations, set-up, and one year of locally offered postharvest training programs by guest instructors. A variety of adaptive research studies were funded but not completed as planned.

The postharvest demonstrations were of relatively low cost, and many could be used over and over again for training on site. When the 14 instructors were asked to estimate the costs for the postharvest demonstrations they had utilized during their training programs, they provided the following information:

- Use of shade—low cost ($50 to $100 for materials and labor), could use any type of local materials to make thatch or a woven roofing/poles structure

- Gentle handling—very low cost, mostly show and tell, $30 for a commercially purchased harvesting bag, much less ($5 to $6) to make one locally

- Use of maturity indices (color charts, sizes)—very low cost (for making color copies, lamination, strong wire to make sizing rings)

- Improved containers (liners, cushions, crates, *etc.*)—very low cost (a few cents for a paper liner, $5 to $7 for a plastic crate)

- ZECC—$400 to $500 for bricks, sand, shade covering, labor, and water tank to construct a new Zero Energy Cool Chamber

- Solar drying practices—$300 to $400 for materials and labor to construct a new direct-style solar drier with six large trays

5. CONCLUSIONS AND RECOMMENDATIONS

5.1. Description of Outcomes and Benefits of PTSC Extension/Outreach Results

Local capacity building in postharvest knowledge, skills, and training expertise in seven countries in SSA was one of the positive outcomes of the project. The follow-on training activities reported on by the 36 ToT participants were able to reach a wide range of local clientele in their home countries, who gained enough knowledge to make changes in their postharvest practices. In Tanzania, the random cluster sampling of 50 farmers, food processors, and marketers who had participated in more than one dozen local training programs offered by the PTSC reported having many positive outcomes. 100% of those surveyed reported that they reduced produce wastage, and 42 of the 50 persons were able to provide details on the costs and positive financial benefits of their changes in postharvest practices. Table 4 provides four examples, and many more examples will be fully documented and published upon completion of the project evaluation for USAID/TOPS.

Table 4. Examples of relative costs and financial benefits of improved postharvest practices for fresh produce as reported in Tanzania.

Interviewee No., Sex, Age, Job Type, Site	Crop and Quantity, Traditional vs. New Practices	Relative Cost, % Losses, Market Value Using Traditional Practice	Relative Cost, % Losses, Market Value Using New Practice	Changes in Income per Load	ROI
#12, male, 52, farmer/processor, Nshupu	Tomatoes, 7600 kg. Selling without grading vs. sorting/grading before selling	0 Tsh 40% 2,850,000 Tsh	160,000 Tsh 10% 3,600,000 Tsh	+590,000 Tsh (US $327) per 7600 kg	Immediate
#20, male, 54, farmer/marketer, Kindi	Cucumbers, 150 kg. Selling without sorting/grading vs. gentle harvest, sorting and grading before selling	0 Tsh 20% 16,000 Tsh	3000 Tsh 5% 30,000 Tsh	+11,000 Tsh (US$ 6.11) per 150 kg	Immediate
#48, female, 31, farmer/processor/marketer, Poli-Ndam	Chinese cabbage, 100 kg. Selling without grading vs. grading before selling	0 Tsh 20% 20,000 Tsh	2000 Tsh 5% 35,000 Tsh	+13,000 Tsh (US$ 7.22) per 100 kg	Immediate
#49, female, 45, farmer/processor, Nshupu	African nightshade, 10 kg. Harvesting under full sun vs. harvesting in morning when temperature is lower	0 Tsh (did not consider her labor to be a cost) 50% 10,000 Tsh	No added cost (her labor only) 5% 20,000 Tsh	+10,000 Tsh (US$ 5.56) per 10 kg	Immediate

Each of the cost-benefit examples in Table 4 were individual case studies based upon recall information on key aspects of the specific technologies being adopted (*i.e.*, on cost of materials and supplies, market prices, percentage of postharvest losses). This detailed information was provided verbally to the interviewers during site visits undertaken for the project evaluation, and the PI performed the calculations based upon the information provided, in order to determine whether there was any relative gain in earnings. A simplified cost-benefit analyses method, developed for the Bill and Melinda Gates Foundation's Appropriate Postharvest Technologies Planning Project, which focuses on measuring relative changes in costs and relative benefits such as changes in percentage of losses and market value per kg was used to make the calculations [13]. All the examples provided by the 42 respondents showed a positive and relatively rapid or immediate return on investment (ROI), since the increase in their earnings using the improved postharvest practice was higher than their initial monetary investment. The change in percentage of losses shown inTable 4 is the key reason for these positive results, as losses using traditional practices ranged from 20% to 50%, and losses using the new practices were reported to be 5% to 10%.

The eight males who were interviewed tended to be involved with handling relatively larger quantities of produce (40 to 7600 kg) and in marketing fresh produce. The 42 females who were interviewed for this evaluation study tended to be involved in food processing of fruits and vegetables, which were sometimes purchased in the wholesale market, and in the handling, processing, and marketing of relatively smaller quantities of produce (5 to 150 kg).

The PTSC model has been observed by many international visitors to the AVRDC site, and was included as a case study on reducing postharvest food losses in Africa in a recent publication by the Global Knowledge Initiative (Rockefeller Fdn/GKI, 2014 [16]). Modified versions of these postharvest training programs are currently being implemented in rural India [17].

5.2. Identification of Constraints and Implementation Issues/Concerns

Key administrators and PTSC managers were interviewed by a WFLO consultant, who met them in their offices and spent a few hours with each person in Arusha during 23–27 February 2015. Seven people were interviewed using a written questionnaire as a guide, followed by probing for specifics and details.

Each person provided information on the constraints they had encountered during the initial PTSC set-up, the implementation of training programs and planned services, and in general management. Most of the constraints had to do with the lack of a budget for operating the PTSC once the Hort CRSP project funding ended in 2013. Several constraints had to do with the limited scope of the mission and operating rules for both AVRDC (which, as a registered NGO, is not allowed to sell goods in Tanzania) and the Ministry of Agriculture, Food Security and Cooperatives (MAFC) (which can offer services and goods for sale at cost, but is generally not allowed to price goods and services in order to make a profit).

In general, the administrators and managers believe the PTSC project to be a success, since they believe that having the PTSC enabled them to increase awareness (among farmers, food processors, marketers, visiting scientists, extension workers, and policy makers) of the role of postharvest technology in reducing food losses. Several people mentioned that the model PTSC was already being copied by other organizations, and that the SARI Agriculture Technology Transfer

Centre site in Njiro had been selected as one of the sites for a "value addition/postharvest training center" under the Market Infrastructure, Value Addition and Rural Finance (MIVARF) project where the retail shop could be managed by a private firm and upgraded to better serve its intended role.

Feedback from participants and formal evaluation results provide guidance on general best practices for management, overcoming difficulties, and making improvements for future postharvest training and services centers.

The PTSC model is already being emulated by several Tanzanian organizations, and a variety of "Value addition centers", "Farmer services centers", "Postharvest training centers", and "Packinghouses/ postharvest training venues" are currently being developed in 16 districts under programs being implemented by MAFC and the Prime Minister's Office. These include:

- a packinghouse for 3500 members of a vegetable cooperative in Lushoto named LUKOVEG (MAFC and the local governmental authority)
- a large citrus/mango packinghouse/training center near Dar es Salaam for a farmers' association of 2000 members (MAFC)
- the Prime Minister's Office/African Development Bank's Market Infrastructure, Value Addition and Rural Finance (MIVARF) project with postharvest training and value addition centers in 12 districts in 2013–2014, with many more planned for 2014–2015, and
- MAFC/Tanzania Horticultural Association (TAHA) Farmer Services Centers (FSCs) under construction in four districts in southern Tanzania and Zanzibar, plus a plan for a large packinghouse to be located near the coast north of Dar es Salaam.

The MIVARF project followed a similar design plan as the Hort CRSP PTSC pilot project, starting with local needs assessments and commodity systems assessments, then working with stakeholders to select key crops and design-appropriate training programs, select and procure equipment, and provide local training for farmers, food processors, and marketers. The MIVARF project has already provided one year of capacity building via ToT programs, and identified local "service providers" in each district who will serve as private sector partners. These established partners are better able to operate the needed postharvest retail shops, provide maintenance, marketing support, and other postharvest services that will help to add value, reduce produce wastage, improve incomes, and create new local businesses and new local jobs. The MIVARF project's postharvest training and services centers will therefore include all five of the key components of the PTSC model.

ACKNOWLEDGMENTS

The authors thank Lizanne Wheeler for her many contributions during the implementation of both the HORT CRSP–funded pilot PTSC project and the TOPS-funded project evaluation. We thank our colleagues at the WFLO International Programs Office for their administrative support.

AUTHOR CONTRIBUTIONS

Lisa Kitinoja conceived and designed the Hort CRSP pilot PTSC project; Diane M. Barrett and Lisa Kitinoja implemented the project as co-PIs; Kitinoja designed the evaluation project for WFLO and Save the Children/TOPS, implemented the evaluation project, and analyzed the data; both Kitinoja and Barrett wrote the paper.

CONFLICTS OF INTEREST

The authors declare no conflict of interest.

REFERENCES

1. Kader, A.A.; Rolle, R.S. *The Role of Post-Harvest Management in Assuring the Quality and Safety of Horticultural Crops*; FAO Agricultural Services Bulletin No.152; Food and Agriculture Organization of the United Nations: Rome, Italy, 2004.

2. Weinberger, K.; Lumpkin, A.T. Diversification into horticulture and poverty reduction: A research agenda. *World Dev.* **2007**, *35*, 1464–1480.

3. AiDA. AidData. Available online: http://aida.developmentgateway.org/aida/Search.do (accessed on 15 January 2010).

4. USAID Documents. Available online: http://dec.usaid.gov/index.cfm?p=search.sqlSearch&CFID=9413138&CFTOKEN=51188616 (accessed on 10 January 2010).

5. World Bank, Projects and Operations. Available online: http://web.worldbank.org/ (accessed on 4 January 2010).

6. UN FAO. INPhO. Available online: http://www.fao.org/inpho/isma?i=INPhO&p=index.jsp&lang=en(accessed on 10 January 2010).

7. DEVEX. Available online: http://www.devex.com/ (accessed on 15 January 2010).

8. Kitinoja, L.; Saran, S.; Roy, S.K.; Kader, A.A. Postharvest technology for developing countries: Challenges and opportunities in research, outreach and advocacy. *J. Sci. Food Agric.* **2011**, *91*, 597–603.

9. Gustavsson, J.; Cederberg, C.; Sonesson, U.; van Otterdijk, R. *Global Food Loss and Food Waste—Extent, Causes and Prevention*; Food and Agriculture Organization of the United Nations: Rome, Italy, 2011.

10. Lipinski, B.; Hanson, C.; Lomax, J.; Kitinoja, L.; Waite, R.; Searchinger, T. *Creating a Sustainable Food Future—Reducing Food Loss and Waste*; WRI Working Paper 39p; World Resources Institute: Washington, DC, USA, 2013.

11. Kitinoja, L.; AlHassan, H.A. Identification of appropriate postharvest technologies for improving market access and incomes for small horticultural farmers in Sub-Saharan Africa and South Asia. Part 1: Postharvest losses and quality assessments. *Acta Hortic.* **2012**, *934*, 31–40.

12. Diem, K.G. Using Research Methods to Evaluate Your Extension Program. Available online: http://www.joe.org/joe/2002december/index.php (accessed on 15 August 2014).

13. Saran, S.; Roy, S.K.; Kitinoja, L. Appropriate postharvest technologies for improving market access and incomes for small horticultural farmers in Sub-Saharan Africa and South Asia. Part 2: Field trial results and identification of research needs for selected crops. *Acta Hortic.* **2012**, *934*, 41–52.

14. Kitinoja, L; Thompson, J.F. Pre-cooling systems for small-scale producers. *Stewart Postharvest Rev.***2010**, *6*, 1–14.

15. Winrock International. *Empowering Agriculture, Energy Options for Horticulture*; United States Agency for International Development: Washington, DC, USA, 2009; p. 79. Available online: http://pdf.usaid.gov/pdf_docs/PNADO634.pdf (accessed on 15 March 2015).

16. Rockefeller Foundation Global Knowledge Initiative. Reducing Food Waste and Spoilage: Assessing Resources Needed and Available to Reduce Post Harvest Food Loss in Africa. 2014. Available online:http://postharvest.org/Rockefeller%20Foundation%20Food%20Waste%20and%20Spoilage%20initiative%20Resource%20Assessment_GKI.pdf (accessed on 2 April 2015).

17. Raman, N.L.M.; Dubey, N. Panchayat and economic empowerment of rural women by hands on training. *Am. Int. J. Res. Humanity Arts Soc. Sci.* **2014**, *5*, 249–252.

4

THE EFFECT OF HIGH DAY AND LOW NIGHT TEMPERATURE ON POLLEN PRODUCTION, POLLEN GERMINATION AND POSTHARVEST QUALITY OF TOMATOES

B. Khanal, A. Suthaparan, A. B. Hückstädt, A. B. Wold, Leiv Mortensen, H. R. Gislerød

Department of Plant and Environmental Sciences, The Norwegian University of Life Sciences, Ås, Norway

ABSTRACT

Temperature integration where high day temperatures are compensated by lower night temperatures is one strategy that can be used to reduce energy consumption in greenhouses. Crop tolerance to temperature variation is a prerequisite for using such a strategy. Greenhouse experiments were conducted on tomatoes cvs, Capricia, Mecano and Cederico in order to investigate the effect of different day/night temperature regimes (24/17, 27/14 and 30/11°C) where the same mean temperature was maintained for the production and germination of pollen. In addition, fruit quality as determined by fruit firmness, dry matter content, soluble solids, titratable acids, and pH was examined at harvest and after seven and 14 days of storage. The 30/11°C treatment significantly increased pollen production and germination compared to the 24/17°C treatment, while the 27/14°C treatment was generally in between the other two treatments. Fruits grown at the 27/14°C treatment were significantly firmer, while fruits grown at 24/17°C had higher dry matter content, soluble solids, and titratable acids compared to the other treatments. There were significant differences between cultivars with respect to firmness, dry matter, titratable acidity, and pH. The quality of the fruits changed during storage, but the storability of the tomatoes was not affected by preharvest temperature treatments. The overall conclusion was that the 27/14°C treatment was superior to the other two temperature treatments with respect to the studied parameters.

Keywords: Daily Mean Temperature; Day Temperature (DT); Night Temperature (NT); Pollen; Temperature Integration; Dry Matter; pH; Titratable Acids; Soluble Solids; Postharvest; Tomato

1. INTRODUCTION

Temperature integration where high day temperatures are compensated by lower night temperatures is an important means of reducing energy consumption in greenhouses [1]. Temperature increases with increasing irradiance to a maximum accepted level at which ventilation takes place, and the night temperature is reduced sufficiently to secure an optimal mean temperature level. Depending on the season and weather conditions, energy savings of more than 20% can be achieved by means of such temperature control [2,3]. However, the effectiveness of temperature integration depends on the plant's ability to tolerate temperature variation. Results of studies conducted on a wide range of vegetables including tomato and ornamentals, have shown that within a certain temperature range, growth and development respond to the mean daily temperature rather than to the day/night temperature variation [4- 6].

Tomato is an important vegetable worldwide, in which temperature is known to affect various physiological aspects, including pollen viability and fruit quality [7]. It has been reported that pollen germination was significantly reduced when tomatoes were grown at temperatures of up to 32/26°C day/night temperature [8]. Tomato quality includes visual characteristics such as color and firmness, nutritional constituents, and organoleptic characteristics such as aroma compounds and the content of sugar and acids. Dorais et al. (2001) reviewed the quality of greenhouse tomatoes [9], and the fruit quality might be affected by high and low temperatures as well as the differences between day and night temperatures [10]. The effect of preharvest and postharvest factors, including temperature, on soluble solids in tomatoes has been reviewed [11]. For cherry tomatoes, the percentage dry weight, glucose, and fructose were found to be higher in fruits developed under high temperature variation (30/

15°C day/night), while lower levels of citric and malic acid were reported [12]. Nevertheless, few works have included the effect of preharvest factors on the quality and storability of tomatoes [13]. It has been concluded that tomato plants can tolerate air temperature fluctuations of up to 6°C from the daily mean of 18.7°C with respect to growth and flowering [14]. Little is known about the effect of temperature variation, with a fixed daily mean temperature, on pollen growth, fruit quality, and storability. The aim of this study was therefore to examine the effect of different day/night temperature variations on pollen production and germination as well as on the postharvest fruit quality of three tomato cultivars.

2. MATERIALS AND METHODS

2.1. Plant Material and Environmental Conditions

Seeds of tomato (Solanum lycopersicum "Capricia", "Mecano" and "Cederico") were sown in 12-cm plastic pots filled with peat (VEKSTTORV, Ullensaker Almenning, Nordkisa, Norway) and perlite (Substraat, RHD, The Netherlands) mixture (3:1). Plants were irrigated with a complete nutrient solution (Superba™ Red (50%) and Calcinite (50%), Yara AS, Oslo, Norway) as necessary. Temperature (day and night) and relative air humidity (RH) were set to 20°C (a mean of 19.7°C observed) and 75% (a mean of 72% observed). Supplemental lighting at a photon flux density (PFD) of 100 μmol $m^{-2}\cdot s^{-1}$ was provided by high-pressure sodium lamps (HPS) (Lucalox LU400/XO/T/40, GE lighting, Budapest, Hungary) when the outside level fell below 100 $W\cdot m^{-2}$global radiation, to maintain 20-hour day length. The PFD was measured by a quantum meter Model QMSW-SS (Apogee instruments Inc., Logan, UT, USA). Once the tomato plants had developed five leaves, they were transplanted into 30-liter plastic pots filled with a peat and perlite mixture (7:3). At the appearance of the first inflorescence, the plants were moved to the experimental conditions.

The experiment was conducted at the Norwegian University of Life Sciences (UMB), Ås, Norway (59° 40'N and 10° 46'E) from March to July 2011, in three greenhouse compartments located beside each other. The ventilation temperature for the different compartments was 24°C, 27°C, and 30°C during the day, and ventilation was used at night to reach temperatures of 17°C, 14°C, and 11°C, respectively. Solar radiation, air temperature, and RH in the greenhouse compartments were recorded at five-minute intervals by a Priva greenhouse computer (Priva, Zijlweg, The Netherlands). The hourly mean values (mean of 12 readings) of these variables were used to calculate the mean day, night, and daily temperatures.

During the experimental period, the mean day temperatures in the three compartments were 23.7°C ± 1.8°C, 26.4°C ± 1.3°C, and 29.2°C ± 2.0°C, and the mean night temperatures were 16.4°C ± 0.1°C, 13.9°C ± 0.1°C, and 11.5°C ± 0.5°C, respectively. The temperature, air humidity, and CO_2 concentration were measured at plant level. During the experimental period, the day length varied from 11 hours in March to 19 hours in June, and no supplementary lighting was used. At sunrise, the vents were closed and the temperature was allowed to increase by means of solar radiation. Night temperatures were achieved by natural cooling through ventilation or by heating. The air humidity was maintained at 75% ± 5% day and night through ventilation or by the application of mist. The CO_2 concentration (as measured every ten minutes) by a Priva infrared gas analyzer) was set to 700 $\mu mol\cdot mol^{-1}$ during the light period when the vents were closed, and to 385 $\mu mol\cdot mol^{-1}$ during ventilation or during the dark period. The mean concentration for the whole experimental period was 628 ± 82, 662 ± 77, and 715 ± 76 $\mu mol\cdot mol^{-1}$ at the 24/17°C, 27/14°C, and 30/11°C treatments, respectively. The temperature and CO_2 concentration during a period of six or 12 days at the time of flowering are summarized (**Table 1**) as well as the temperature during the four weeks prior to harvest (**Table 2**). The photosynthetic active radiation (PAR) inside the greenhouse (about 50% of the outside radiation) during the 12 days was 13.3 ± 3.1 (17 - 28 March) and 12.2 ± 6.4 mol $m^{-2} \cdot day^{-1}$ (29 March-10 April) for truss number four and six, respectively. The photosynthetic active radiation was 17 ± 6, 20 ± 6 and 20 ± 10 $mol \cdot m^{-2} \cdot day^{-1}$ in April, May and June, respectively. The conversion factor from global radiation (in $MJ \cdot m^{-2} \cdot day^{-1}$) to PAR (in $mol \cdot m^{-2} \cdot day^{-1}$) was 2.2.

Elemental sulfur was applied for two hours every night. Plants were fertilized daily with four liters (complete saturation of the substrate) of the same nutrient solution as previously described. The nutrient solution had a conductivity of 2.5 $mS \cdot cm^{-1}$ (DGT Volmatic Type LM20 Serial 9305), and the salinity of the growing medium was 4 - 5 $mS \cdot cm^{-1}$, using the soil saturated extract (SSE) method.

The plants were pinched above the ninth truss in all temperature treatments. The fourth and sixth trusses as counted from the bottom of the plants were selected for the following analysis: The first flower was assessed for pollen production and the second flower for pollen germination one day after opening, and the third flower for pollen production and the fourth flower for pollen germination four days after opening. This was done under the same conditions (20°C/70% RH) for all three temperature treatments. In addition, the fifth and sixth flowers on the same trusses were used to assess pollen pro duction and germination at their respective growth temperatures. Since the solar radiation is quite variable during spring two different trusses were selected in order to cover a longer time period for climate exposure.

2.2. Pollen Production

In order to measure pollen production, the flowers were removed and placed in 50 ml centrifuge tubes that were filled with 5 ml of distilled water containing 20 $\mu l \cdot l^{-1}$ of Tween 20 surfactant (Sigma-Aldrich Chemie GmbH, Steinheim, Germany) (one flower per tube). Tubes were shaken by hand 40 times and the flower was removed. The number of pollen per ml of suspension was calculated in two aliquots of suspension using haemocytometer (HYCOR, Hycor biomedical inc. California, USA) under a light microscope at 100X. Pollen with a diameter of less than 20 μm and shrunken pollen were considered undersized or abnormal.

Table 1. Day and night temperature (means ± SD) during a six-day period (from five days before until one day after flower opening) or a twelve-day period (from eight days before until four days after flower opening) in trusses four and six. CO_2 concentration is given for the light period.

Set temperature (°C)	Truss no.	Day temp. (°C)		Night temp. (°C)		Mean temp. (°C)		CO_2 conc. (μmol·mol^{-1})
		6-day period	12-day period	6-day period	12-day period	6-day period	12-day period	6-day/12-day period
24/17	4	23.7 ± 1.3	23.9 ± 1.2	16.4 ± 0.1	16.3 ± 0.1	19.1	19.2	632 ± 82/618 ± 80
27/14	4	26.0 ± 1.1	26.4 ± 1.4	13.9 ± 0.1	13.8 ± 0.1	18.2	18.4	630 ± 76/652 ± 91
30/11	4	28.7 ± 2.3	29.1 ± 2.2	11.4 ± 0.2	11.1 ± 0.2	17.5	17.7	652 ± 88/716 ± 94
24/17	6	22.3 ± 2.1	23.5 ± 2.3	16.5 ± 0.1	16.6 ± 0.1	18.9	19.3	624 ± 82/580 ± 82
27/14	6	26.2 ± 2.4	26.4 ± 2.1	13.9 ± 0.1	14.1 ± 0.1	18.4	18.9	694 ± 78/624 ± 95
30/11	6	29.5 ± 2.3	29.3 ± 2.2	11.1 ± 0.2	12.0 ± 0.7	18.1	18.5	778 ± 73/720 ± 92

Table 2. Mean maximum day and minimum night temperature during the four weeks prior to harvesting the tomato fruits of truss number five for three different temperature treatments.

Set day/night temperature (°C)	Observed day temperature (°C)	Observed night temperature (°C)	Mean temperature (°C)
24/17°C	25.4 ± 2.2	15.3 ± 1.4	19.7
27/14°C	27.2 ± 1.3	14.1 ± 0.5	19.4
30/11°C	30.1 ± 2.5	11.3 ± 0.5	19.6

2.3. Pollen Germination

For pollen germination, the flowers were picked with their pedicle and held 2 cm above Petri dishes (5 cm in diameter) containing pollen growth media [15]. Flowers were vibrated for five seconds using an electric tooth brush (Philips HX1610 Double cleaning action, China)

placed on the flower pedicle. Petri dishes were then sealed and incubated in a growth chamber at 20°C and 70% RH. The day length was 14 hours and 130 ± 10 $\mu mol \cdot m^{-2} \cdot s^{-1}$ PFD was supplied by high-pressure mercury lamps (Powerstar HQI-BT 400 W/D day light, OSRAM GmbH, Augsburg, Germany). One day after incubation, a piece of the pollen growth media was cut and placed on a microscopic glass slide. A droplet of water was placed on top and covered with cover slips. The samples were then assessed for pollen germination under a light microscope. Pollen containing germ tubes at least half the length of the diameter of the pollen was deemed to have germinated. The first 100 pollens were assessed and the percentage of pollen germination was calculated.

2.4. Fruit Harvest and Postharvest Quality

Forty-five tomato fruits from the third and fifth clusters of each cultivar and temperature treatment were harvested at commercial ripening stage based on comparison with color chart scale 6 - 9 (Ctifl, Code Couleur Tomate, France). At harvest (day 0), 15 tomatoes were randomly selected and divided into three replicates each containing five tomatoes. The color and firmness of the fruits were measured using color chart comparison and DUROFEL DFT 100 digital firmness tester (Agro-Technologie, St Etienne du Gres, France), respectively. For firmness measurements, the tip of the digital firmness tester was placed on the surface of each tomato fruit at three different points. The instrument measured the elasticity of the outer fruit flesh within the range from 0 (no resistance or high elasticity) to 100 (high resistance or no elasticity) and firmness is specified as DUROFEL-units. The remaining 30 fruits were stored in darkness at 13°C and 85% RH. After seven and 14 days' storage, 15 fruits were assessed for color and firmness. Immediately after the assessments, 3 × 5 fruits were frozen at −20°C pending further analysis.

2.5. Dry Matter Content, Soluble Solids, Titratable Acidity, and pH

For analysis of dry matter, soluble solids, titratable acidity, and pH, frozen tomatoes were thawed overnight at room temperature. Tomato samples were homogenized using a food processor (BRAUN, Germany). For dry matter assessment, six grams of homogenized material was dried at 104°C for 24 hours and the dry matter percentage was calculated.

The homogenized samples were filtered (125 mm, Whatman GmbH, Dassel, Germany) and the filtrates were used for further analysis. For soluble solids measurement, a digital refractometer, Model Atago Palett PR-100 (Atago Co., Ltd., Tokyo, Japan) was first calibrated using distilled water. Then two to three drops of tomato juice were placed on the sensor of the refractometer. Soluble solids were given as a percentage.

Automatic titrator, Model Methrom 716 DMS Titrino and 730 Sample changer (Metrohm Ltd., Herisau, Switzerland) were used for the measurement of titratable acidity. Ten ml of filtrate was diluted with 50 ml of distilled water. Then sodium hydroxide (0.1N) was added to the diluted filtrate to reach pH of 8.1. Titratable acidity was calculated as a percentage of citric acid. The pH was measured using a pH meter (Model 691 PH Meter, Metrohm Ltd., Herisau, Switzerland).

2.6. Statistical Analysis

Minitab (version 16) was used to conduct analysis of variance (GLM procedure). Three different day/night temperature combinations and three different varieties were included in the model. Response variable means were compared using Tukey's pairwise comparison test at P = 0.05.

3. RESULTS

3.1. Pollen Production (Figure 1)

The number of normal pollen in the fourth truss one day after flower opening was not significantly affected by day/night temperature, while pollen production in the sixth truss was significantly higher ($p < 0.05$) at 27/14°C and 30/11°C compared to 24/17°C (**Table 3**). The number of normal pollen assessed four days after flower opening as a mean of the three cultivars, increased as DT increased/NT decreased above/below 24/17°C (**Table 3**). The number of abnormal pollen often increased with increasing DT/decreasing NT, particularly in the 30/11°C treatment (**Table 3**). The number of normal pollen was highest in cv. Cederico as measured one day after flower opening in the fourth truss, and in cv. Mecano as meas ured four days after flower opening in the sixth truss (data not presented).

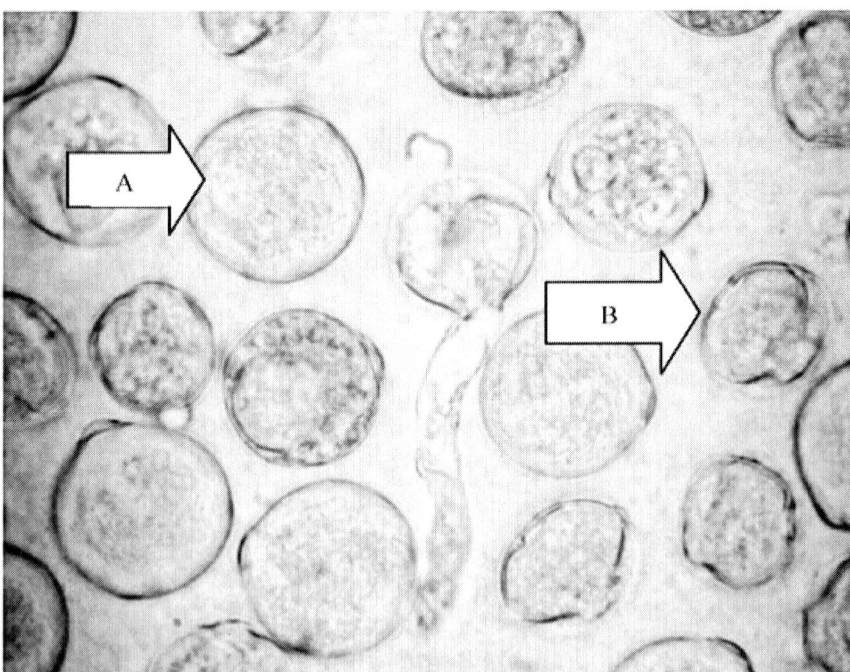

Figure 1. Normal pollen (A) and abnormal pollen (B).

Table 3. Effect of day/night temperature regimes on number of normal and abnormal pollen in different trusses one and four days after flower opening. Values represent means of three cultivars and two replicates (n = 6, ±SD). Different letters within each column indicate significant differences between treatments at $p < 0.05$ level.

Set temperature (°C)	Truss no.	No. of normal pollen		No. of abnormal pollen	
		1 day	4 days	1 day	4 days
24/17	4	28.1 ± 2.0 a	40.3 ± 2.3 b	8.7 ± 0.8 a	15.0± 1.4b
27/14	4	28.2 ± 2.2 a	41.6 ± 2.3 ab	9.1 ± 1.2 a	18.9±1.5 ab
30/11	4	31.3 ± 2.5 a	48.8 ± 1.6 a	9.3 ± 0.8 a	20.5± 1.3 a
24/17	6	26.4 ± 1.4 b	33.4 ± 1.7 b	11.0 ± 1.2 b	6.7± 0.7 c
27/14	6	45.9 ± 2.9 a	42.1 ± 2.1a	9.3 ± 0.9 b	10.6± 0.6 b
30/11	6	51.0 ± 2.1 a	43.4 ± 1.6a	15.3 ± 1.3 a	15.5± 0.7 a

3.2. Pollen Germination

The pollen germination percentage as measured under the same climate conditions (in a growth chamber) generally increased as DT increased/NT decreased (**Table 4**). This was the case in both trusses as well as for pollen harvested one or four days after flower opening (**Table 4**). Similar results were found when pollen germination was measured under growing conditions (**Table 4**). In growth chamber incubation, pollen germination of the cultivar Cederico was significantly higher four days after flower opening in the sixth truss compared to Capricia and Mecano (data not presented).

3.3. Tomato Quality

Tomatoes grown at 27/14°C were significantly firmer compared to the other temperature treatments (**Table 5**). Tomatoes grown at 24/17°C contained significantly higher amounts of dry matter, soluble solids and titratable acidity, whereas pH was higher in tomatoes grown at 30/11°C. The cultivars Capricia and Mecano had significantly firmer fruits compared to Cederico. No significant differences were found between the cultivars with respect to soluble solids. Significant differences were observed between the cultivars with respect to dry matter, titratable acidity, and pH. Fruit firmness decreased significantly during storage. Dry matter, soluble solids, and pH remained stable during storage, whereas titratable acidity decreased significantly from day seven to day 14. Visually observed the tomatoes produced at 30/11°C were pale in color compared to tomatoes produced in the other temperature treatments, however, there was no systematic record of the color. The different preharvest temperature treatments had no significant effect on the storability of the tomato fruits (data not shown).

Table 4. Effect of day/night temperature regimes on percentage pollen germination as measured under the same conditions in a growth chamber and under growing conditions one day and four days after flower opening. Values represent means of three cultivars and two replicates (n = 6, ±SD).

Set temperature (°C)	Truss no.	Same conditions in growth chamber		At growing conditions	
		1 day	4 days	1 day	4 days
24/17	4	26.4 ± 1.8 b	19.1 ± 1.7 b	15.9 ± 1.7 b	15.1 ± 1.5 c
27/14	4	36.5 ± 2.9 a	24.5 ± 2.2 b	17.0 ± 1.2 b	23.6 ± 1.4 b
30/11	4	41.1 ± 3.5 a	34.8 ± 2.8 a	37.0 ± 2.1 a	37.0 ± 2.1 a
24/17	6	16.2 ± 1.2 c	17.3 ± 1.2 c	18.2 ± 2.1 b	17.1 ± 1.5 c
27/14	6	28.6 ± 2.5 b	25.4 ± 1.6 b	24.4 ± 1.1 b	27.8 ± 1.0 b
30/11	6	40.0 ± 2.0 a	37.7 ± 1.9 a	38.1 ± 2.3 a	41.1 ± 2.2 a

4. Discussion

A climate control strategy where high DT was combined with low NT, a DT of at least about 29°C combined with a NT of 11°C - 12°C, did not have a negative effect on the production and germination of normal pollen. Compared to the control treatment of 24/17°C, this type of climate was often actually beneficial to pollen production and germination. This was concluded when the pollen was tested under plant growing conditions as well as under the same climate conditions in a growth chamber. Although the number of abnormal pollen under high DT/low NT increased, this was more than compensated by an increased number of normal pollen and increased germination potential. Sato et al. (2006) found that increasing the DT to 32°C combined with a NT of 26°C had a negative impact on pollen germination in tomato [8]. The maximum DT was higher and the NT much higher than that used in the present experiment. Energy reserves seem to be the predominant factor in determining pollen production as well as pollen viability [16]. Delayed ventilation and higher CO_2 concentration accompanied higher DT, and this probably stimulated the photosynthetic rate. It has been reported that the negative effects of high day temperatures of up to about 30°C on photosynthesis can be minimized or eliminated by CO_2 enrichment in roses [17,18] and cucumber [19]. Photosynthesis has also been found to be stimulated by higher growth temperatures when the CO_2 concentration was increased [20]. The difference in pollen production in the fourth and sixth truss one day after opening might also be due to longer day lengths.

Table 5. Effect of different day/night temperature regimes during growth on firmness, dry matter (%), soluble solids (%), titratable acidity (%) and pH of three tomato cultivars at harvest and after seven and fourteen days of storage (n = 6, ±SD).

Treatments		Firmness	Dry matter (%)	Soluble solids (%)	Titratable acid (%)	pH
Day/Night temperature	24/17°C	83.6 ± 0.8 b	5.55 ± 0.04 a	5.12 ± 0.04 a	0.450 ± 0.006 a	4.18 ± 0.01 b
	27/14°C	86.5 ± 0.7 a	4.88 ± 0.05 b	4.64 ± 0.02 b	0.426 ± 0.004 b	4.18 ± 0.01 b
	30/11°C	81.7 ± 0.8 b	4.89 ± 0.04 b	4.73 ± 0.01 b	0.403 ± 0.003 c	4.23 ± 0.01 a
Cultivars	Capricia	86.5 ± 0.6 a	5.24 ± 0.05 a	4.83 ± 0.05 a	0.438 ± 0.006 a	4.18 ± 0.01 b
	Mecano	84.0 ± 0.5 a	5.11 ± 0.06 ab	4.87 ± 0.04 a	0.422 ± 0.003 ab	4.21 ± 0.00 a
	Cederico	81.2 ± 0.8 b	4.97 ± 0.05 b	4.78 ± 0.03 a	0.418 ± 0.005 b	4.20 ± 0.01 ab
Storage time	0 day	89.1 ± 0.5 a	5.17 ± 0.06 a	4.84 ± 0.04 a	0.427 ± 0.004 ab	4.20 ± 0.01 a
	7 days	84.1 ± 0.5 b	5.14 ± 0.06 a	4.88 ± 0.04 a	0.435 ± 0.006 a	4.19 ± 0.01 a
	14 days	81.7 ± 0.8 c	5.01 ± 0.06 a	4.79 ± 0.04 a	0.417 ± 0.005 b	4.20 ± 0.01 a

Fruits developed at 27/14°C were firmer than fruits developed under other temperature treatments, and this might be related to the temperature-dependent activity of cell wall degrading enzymes acting on proteins and carbohydrates [21]. The difference in firmness between cultivars could be explained by variation in skin toughness, flesh firmness, and the pericarp/locular material ratio [22]. The firmness decreased during storage, which is most probably related to possible water loss and further ripening of the fruit. This is in accordance with the results of Jha and Matsuoka (2005), where a significant reduction in tomato firmness was observed during storage [23]. The red color of tomato fruit is determined by the amount of lycopene present in pericarp [24,25]. The optimal temperature for lycopene synthesis in tomatoes is in the range 16°C - 21°C, and a very high DT (30°C) in the present experiment reduced the red color as visually observed, even if this high DT was compensated by 11°C NT.

The dry matter of fruit and vegetables is mainly composed of sugars and acids. Tomatoes usually contain 5% - 8% dry matter of which 4% - 6% is soluble solids [9]. Citric and malic acids are the main organic acids found in tomato fruits and constitute approximately 10% - 13% of the dry matter content [9]. In the present study, the dry matter content of the tomato fruits that could be associated with reduced soluble solids and titratable acids, decreased with increasing DT/decreasing NT. This reflects a somewhat reduced organoleptic quality of the fruits at high DT/low NT. The dry matter and soluble solids content of fruit mainly depends on the synthesis and transport of assimilates from the leaves to the fruits [26]. High temperatures are known to favor the distribution of assimilates to the fruits during fruit development [27], whereas low night temperatures have previously been shown to reduce the content of soluble solids in tomatoes [28]. For cherry tomatoes grown in greenhouses, an increase in sugars and a decrease in titratable acidity were observed in late harvest when the temperature and solar radiation peaked [12]. During storage, the content of soluble solids remained stable whereas the titratable acidity changed significantly. The stability of soluble solids during storage has previously been observed [29,30] and the concentration of titratable acidity has been found to decrease [30] during a seven-day storage period.

5. CONCLUSION

It can be concluded that significantly larger variations between day and night temperatures than commonly applied will not reduce the pollen production and pollen germination potential in tomato as long as the mean temperature is kept constant. The overall conclusion was that the 27/14°C treatment was superior to the other two temperature treatments (24/17°C and 30/11°C DT/NT). The preharvest temperature regimes did not affect the storability of the tomatoes.

6. ACKNOWLEDGEMENTS

This project was financed by the Research Council of Norway. We would like to thank Ida Kristin Hagen, Dag Wenner, Signe Hansen and Kari Grønnerød for their technical assistance.

REFERENCES

1. S. R. Adams, V. M. Valdes, F. A. Langton and P. J. C. Hamer, "Reducing Carbon Emissions from Greenhouse Production through the Use of Temperature Integration and Alternative Sources of Heat," Acta Horticulturae, Vol. 893, No. 1, 2011, pp. 95-101.

2. N. Sigrimis, A. Anastasiou and N. Rerras, "Energy Saving in Greenhouses Using Temperature Integration: A Simulation Survey," Computers and Electronics in Agriculture, Vol. 26, No. 3, 2000, pp. 321-341.

3. O. Korner, M. J. Bakker and E. Heuvelink, "Daily Temperature Integration: A Simulation Study to Quantify Energy Consumption," Biosystems Engineering, Vol. 87, No. 3, 2004, pp. 333-343.

4. J. C. Bakker and J. A. M. Van Uffelen, "The Effects of Diurnal Temperature Regimes on Growth and Yield of Glasshouse Sweet Pepper," The Netherlands Journal of Agricultural Science, Vol. 36, 1988, pp. 201-208.

5. B. J. Bailey, H. Challa, A. A. Risjdijk and J. V. M. Vogelezang, "Computers and Automation—Temperature Integration on a 24-Hour Base: A More Efficient Climate Control Strategy," Acta Horticulturae, Vol. 519, No. 1, 2000, pp. 163-169.

6. R. G. Hurd and C. J. Graves, "The Influence of Different Temperature Patterns Having the Same Integral on the Earliness and Yield of Tomatoes," Acta Horticulturae, Vol. 148, 1984, pp. 547-554.

7. A. Picken, "A Review of Pollination and Fruit Set in the Tomato (Lycopersiconesculentum Mill.)," The Journal of Horticultural Science & Biotechnology, Vol. 59, 1984, pp. 1-13.

8. S. Sato, M. Kamiyama, T. Iwata, N. Makita, H. Furukawa and H. Ikeda, "Moderate Increase of Mean Daily Temperature Adversely Affects Fruit Set of Lycopersiconesculentum by Disrupting Specific Physiological Processes in Male Reproductive Development," Annals of Botany, Vol. 97, No. 5, 2006, pp. 731-738.

9. M. Dorais, A. P. Papadopoulos and A. Gosselin, "Greenhouse Tomato Fruit Quality," In: J. Janick, Ed., Horticultural Reviews, John Wiley & Sons, Inc., Oxford, 2010.

10. N. Gruda, "Impact of Environmental Factors on Product Quality of Greenhouse Vegetables for Fresh Consumption," Critical Review in Plant Science, Vol. 24, No. 3, 2005, pp. 227-247.

11. D. M. Beckles, "Factors Affecting the Postharvest Soluble Solids and Sugar Content of Tomato (Solanumlycopersicum L.) Fruit," Postharvest Biology and Technology, Vol. 63, No. 1, 2012, pp. 129-140.

12. M. A. Rosales, L. M. Cervilla, E. Sanchez-Rodriguez, M. Rubio-Wilhelmi, B. Blasco, J. J. Rios, T. Soriano, N. Castilla, L. Romero and J. M. Ruiz, "The Effect of Environmental Conditions on Nutritional Quality of Cherry Tomato Fruits: Evaluation of Two Experimental Mediterranean Greenhouses," Journal of the Science of Food and Agriculture, Vol. 91, No. 1, 2011, pp. 152-162.

13. M. Linke and H. P. Kläring, "Effect of Different Preharvest Conditions on the Postharvest Keeping Quality of Greenhouse Tomatoes," Acta Horticulturae, Vol. 654, 2004, pp. 213-219.

14. A. N. M. De Koning, "Long-Term Temperature Integration of Tomato: Growth and Development under Alternating Temperature Regimes," Scientia Horticulturae, Vol. 45, No. 1-2, 1990, pp. 117-127. J. L. Brewbakerand B. H. Kwack, "The Essential Role of Calcium Ion in Pollen Germination and Pollen Tube Growth," American Journal of Botany, Vol. 50, No. 9, 1963, pp. 859-865.

15. N. Firon, R. Shaked, M. M. Peet, D. M. Pharr, E. Zamski, K. Rosenfeld, L. Althan and E. Pressman, "Pollen Grains of Heat Tolerant Tomato Cultivars Retain Higher Carbohydrate Concentration under Heat Stress Conditions," Scientia Horticulturae, Vol. 109, No. 3, 2006, pp. 212-217.

16. D. R. Taub, J. R. Seemann and J. S. Coleman, "Growth in Elevated CO_2 Protects Photosynthesis against High-Temperature Damage," Plant, Cell & Environment, Vol. 23, No. 6, 2000, pp. 649-656.

17. L. M. Mortensen and H. R. Gislerød, "The Effect of High CO_2 Concentrations on Diurnal Photosynthesis at High Daytime Temperatures in Small Stands of Cut Roses," European Journal of Horticultural Sciences, Vol. 77, No. 4, 2012, pp. 163-169.

18. L. M. Mortensen and H. R. Gislerød, "The Effect of CO_2 Concentration on the CO_2Exchange Rate in a Small Stand of Cucumber during Different Periods of the Day," European Journal of Horticultural Sciences, Vol. 77, No. 1, 2012, pp. 24-30.

19. R. Besford, L. Ludwig and A. Withers, "The Greenhouse Effect: Acclimation of Tomato Plants Growing in High CO_2, Photosynthesis and Ribulose-1, 5-Bisphosphate Carboxylase Protein," Journal of Experimental Botany, Vol. 41, No. 8, 1990, pp. 925-931.

20. V. S. Meli, S. Ghosh, T. N. Prabha, N. Chakraborty, S. Chakraborty and A. Datta, "Enhancement of Fruit Shelf Life by Suppressing N-Glycan Processing Enzymes," Proceedings for National Academy of Science, Vol. 107, No. 6, 2010, pp. 2413-2418.

21. D. M. Barret, E. Garcia and J. E. Wayne, "Textural Modification of Processing Tomatoes," Critical Reviews in Food Science and Nutrition, Vol. 38, No. 3, 1998, pp. 173- 258.

22. S. N. Jha and T. Matsuoka, "Determination of Post-Harvest Storage Life of Tomato Fruits," Journal of Food Science and Technology, Vol. 42, No. 6, 2005, pp. 526-529.

23. R. Arias, T. C. Lee, L. Logendra and H. Janes, "Correlation of Lycopene Measeured by HPLC with the L*, a*, b* Color Readings of a Hydroponic Tomato and the Relationship of Maturity with Colour and Lycopene Content," Journal of Agricultural Food Chemistry, Vol. 48, No. 5, 2000, pp. 1697-1702.

24. S. L. Molyneux, C. E. Lister and G. P. Savage, "An Investigation of the Antioxidant Properties and Colour of Glasshouse Grown Tomatoes," International Journal of Food Sciences and Nutrition, Vol. 55, No. 7, 2004, pp. 537-545.

25. P. Riga, M. Anza and C. Garbisu, "Tomato Quality Is More Dependent on Temperature than on Photosynthetically Active Radiation," Journal of the Science of Food and Agriculture, Vol. 88, No. 1, 2008, pp. 158-166.

26. A. N. M. De Koning, "The Effect of Temperature on Fruit Growth and Fruit Load of Tomato," Acta Horticulturae, Vol. 248, 1989, pp. 329-336.

27. H. Qi, L. Hua, L. Zhao and L. Zhou, "Carbohydrate Metabolism in Tomato (Lycopersicaonesculentum Mill.) Seedlings and Yield and Fruit Quality as Affected by Low Night Temperature and Subsequent Recovery," African Journal of Biotechnology, Vol. 10, No. 30, 2011, pp. 5743- 5749.

28. J. Javanmardi and C. Kubota, "Variation of Lycopene, Antioxidant Activity, Total Soluble Solids and Weight Loss of Tomato during Postharvest Storage," Postharvest Biology and Technology, Vol. 41, No. 2, 2006, pp. 151- 155.

29. H. Auerswald, P. Peters, B. Bruckner, A. Krumbein and R. Kuchenbuch, "Sensory Analysis and Instrumental Measurement of Short-Term Stored Tomatoes (Lycopersiconesculentum Mill.)," Postharvest Biology and Technology, Vol. 15, No. 3, 1999, pp. 323-334.

5

EFFECT OF ORGANIC PRODUCTION SYSTEMS ON QUALITY AND POSTHARVEST PERFORMANCE OF HORTICULTURAL PRODUCE

Francesco Giovanni Ceglie[1,2], Maria Luisa Amodio[2] and Giancarlo Colelli [2]

[1]Organic Agriculture Department, Mediterranean Agronomic Institute of Bari, CIHEAM-IAMB, Via Ceglie, 9, Valenzano, BA 70010, Italy

[2]Department of Science of Agriculture, Food and Environment, University of Foggia, Via Napoli 25, Foggia, FG 71122, Italy

ABSTRACT

Organic standards include a well-defined set of practices and a list of technical tools that are permitted by regulation. Organic products are mainly purchased for their safety and absence of synthetic pesticide residues. Furthermore, a diet based on organic products claims to provide health benefits due to the high nutritional value compounds that are more concentrated in organic products compared to conventional ones. As the scientific basis of the differences between organically- and conventionally-grown fruits and vegetables is under debate, some of the published work, together with some recent unpublished results, will be covered in the present review. In addition, the effect of different approaches to organic horticultural production will be described. Many studies have confirmed lower nitrate content, especially in leafy vegetables, and higher antioxidant compounds in organically-grown fruits in comparison to conventional ones. A recent study reported organic kiwifruit as higher in ascorbic acid and total phenol content than conventional kiwifruit. These differences were maintained throughout cold storage. Similarly, in organic grapes, antioxidant-related compounds were significantly higher than in conventionally-grown grapes. Analogous results were obtained with organic strawberries grown in protected conditions. However, conventional products usually result in higher moisture content, and this should be taken into account to confirm the differences on a dry matter basis. Possible explanations for the effects of organic farming practices on nutritional quality and postharvest performance of fresh produce are the following: (i) organic amendments provide a high input of exogenous organic matter and of nutrients for a long period; in contrast, mineral fertilizers, allowed only in conventional farming systems, are highly concentrated in nutrients that are directly available for root uptake in a shorter time period; (ii) the use of synthetic pesticides (only possible in conventional agriculture) slows down defence mechanisms against pathogens, with the consequence of favoring primary metabolism; (iii) cultural practices may result in different plant composition and nutritional quality, which in turn influence cold storage performance of the products as these differences, both in fertility and pest management, affect the allocation of secondary plant metabolites (such as ascorbic acid and phenolic compounds).

Keywords: shelf life; secondary metabolites; oxidative stress; strawberry; tomato; nutrition

1. INTRODUCTION

Organic standards include a well-defined set of practices and a list of technical tools that are permitted by regulations (*i.e.*, Reg n.889/08 in UE and the National Organic Program in U.S.). A diet based on organic products claims to provide health benefits due to the higher concentration of nutritional compounds compared to conventional ones, and the absence of pesticide residues [1]. The present challenge of feeding the world requires new strategies to ensure food security which is surely based on food availability and access, but also on food safety and nutritional quality. Organic production systems may be a way to ensure the sustainability of production, allowing preservation of natural resources for present and future generations, while providing a high quality and long shelf life of the product [2]. Despite the importance of this issue, few comparative studies have focused on postharvest aspects, and the scientific basis of the possible difference between organically- and conventionally-managed produce is still debated [3]. Some published studies, together with some recent unpublished results, will be covered in the present review. In particular, comparison studies that were conducted under similar environmental conditions with respect to climate, soil characteristics and availability of nutrients, have been considered for the effect of diverse pre-harvest practices on postharvest performance.

2. DIVERSE PLANT RESPONSE TO DIFFERENT SYSTEMS OF PRODUCTION

The overall quality characteristics of the plant parts which are used as food are the result of the interactions among genotype, environmental conditions, cultural practices and postharvest handling and processing techniques. Any environmental factor that varies from the ideal potentially generates a stress which may either promote or inhibit specific plant responses. Such stress may lead to oxidative stress that produces reactive oxygen species (ROS) in the chloroplasts (free radicals like superoxide and molecular forms like singlet oxygen) and in the mitochondria (mainly superoxide). Ascorbic acid, glutathione, phenolic compounds, and alkaloids, among the antioxidants, are non-enzymatic defenses used by the plant to cope with this dangerous situation [4]. In general, an organic production system is a stressful system largely due to an insufficient supply of mineral nitrogen throughout the crop cycle, fostering phenolics and ROS which are natural defense substances. For organic plants, the content of secondary metabolites, that also include antioxidants, have been positively correlated with these natural defense substances [5]. However, antioxidant biosynthesis requires plant resource reallocation from primary metabolism to secondary metabolite production. As a consequence, cultural practices, both organic and conventional, may result in different plant composition and nutritional quality at harvest due to different biotic and/or abiotic stress conditions, which, in turn, influence the concentration of secondary metabolites.

3. QUALITY AND POSTHARVEST PERFORMANCES OF ORGANIC AND CONVENTIONAL FRUIT AND VEGETABLES

In the scientific literature, different studies have claimed opposite results. Woëse *et al.* [6] and Smith Spangler *et al.* [3] estimated large but not statistically significant differences in quality between organic and conventionally-managed crops, which led them to conclude that the two production systems had similar qualitative characteristics. However, in the last decade, other studies have contributed to enlarge the dataset for this comparison. Lairon [7] confirmed the absence of pesticide residues in organic food (97% of samples had no detectable level of pesticide) and reported a lower content of nitrogen in organic vegetables in comparison to conventional ones. Similarly, Baranski *et al.* [8] concluded that the concentration of total nitrogen was 10% lower in organic compared to conventional crops (nitrate 30% and nitrite 87% lower) [9]. The low percentage of nitrogen forms was correlated with a high concentration of secondary metabolites (such as phenols and vitamins, which do not contain nitrogen). Several studies have confirmed elevated levels of secondary metabolites in organic carrots [10], sweet peppers [11], and tomatoes [12]. Other studies have reported opposite results with tomatoes [13] and carrots [14]. Plant secondary metabolites are important for their role in enhancing human health [15], and, as a result, much attention has been dedicated to this family of compounds in the recent comparison studies. Luthria [16] found similar phenolic acid content in eggplant cultivated with organic or conventional techniques. Chassy *et al.* [17] reported higher values of antioxidant activity in organic tomato and bell peppers, while D'Evoli *et al.* [18] found that total polyphenol content was higher in conventional plums, but ascorbic acid, α-tocoferol, and β-carotene were higher in organic samples. Furthermore, the majority of studies comparing the polyphenol content of plant products from different farming systems indicated significantly higher concentrations of these substances in organic fruits and vegetables compared to conventional ones. Lairon [7] highlighted the high level of antioxidants, minerals, and dry matter in organically-grown products. However, conventional products usually result in higher water content, and this should be taken into account with the differences confirmed in terms of dry matter. Bourn and Prescott [19] observed higher dry matter content in organically-grown carrots and spinach in comparison with the conventional ones. Ceglie *et al.* [20] and Conti *et al.* [21] found a higher dry matter content in organic strawberries. Dry matter at harvest has been proposed as a promising predictor of post-harvest soluble solids in apples, also affecting the relative amount of mass that could be lost during cold storage [22]. Woëse *et al.* [6] reported no clear trend in the level of total sugar content in organic vegetables compared with conventional ones.

Different types and amounts of secondary metabolites at harvest may affect antioxidant changes during the cold storage period and determine the diverse postharvest performance of the products. Despite the relevance of the issue, few comparative studies have focused on such postharvest issues. Differences during cold storage in incidence of physiological disorders, soluble solids content, firmness, and mineral content have been reported [23,24]. After ethylene treatment, organic banana showed faster peel color changes, and lower gravimetric water balance and pulp/peel ratio, and impedance in comparison to conventionally-grown banana [25]. During simulated marketing conditions, organoleptic characteristics and resistance to deterioration were higher for organic strawberries in comparison to the conventional fruit [26]. Hasey *et al.*[27] reported higher soluble solids content and firmness in organically-grown kiwifruit compared with conventionally-grown kiwifruit. In contrast, Benge *et al.* [28] reported that conventional kiwifruit showed higher soluble solids content and similar softening behavior and decay than organic ones, analyzed at the same firmness stage. The levels of calcium in kiwifruit were negatively correlated with the incidence of soft patches. Moreover, higher levels of Ca as well as of NO_3^-, Mg, Fe, and Zn were observed at harvest and during 25 days of cold storage at 10 °C for conventional "Meyer" lemons (*Citrus meyeri* Tan.) in comparison with fruits from an organic orchard [29].

4. POSTHARVEST PERFORMANCE OF ORGANICALLY AND CONVENTIONALLY GROWN KIWIFRUITS

Amodio *et al.* [30] compared postharvest performance of organic and conventional "Hayward" kiwifruit harvested at the same maturity stage. Fruit shape and peel characteristics before cold storage, maturity indices (CO_2 and C_2H_4 production, firmness, color, soluble solids content, and acidity) and compounds associated with flavor and nutritional quality (minerals, sugars and organic acids, ascorbic acid, total phenols, and antioxidant activity) were determined at 0, 35, 72, 90, and 120 days of cold storage at 0 °C, and after 1 week of shelf-life simulation at 20 °C. At harvest, organic and conventional kiwifruit had similar soluble solids content. Conventional kiwifruit had a higher firmness and L* value, and a lower hue angle and chromaticity, resulting in a lighter green color when compared with the organic kiwifruit. During cold storage, soluble solids content increased more in conventional than in organic kiwifruit. Concerning nutritional compounds, ascorbic acid, and total phenols were more concentrated in organic than conventional kiwifruit, resulting in a higher antioxidant activity. These differences were maintained throughout the cold storage period. The two production systems resulted in different morphological attributes since organic kiwifruits exhibited a larger total and columella area, smaller flesh area, more spherical shape, and thicker skin compared to conventional kiwifruit. Also, the levels of the main mineral constituents were higher in organic than in conventional kiwifruit.

5. POSTHARVEST QUALITY OF ORGANIC AND CONVENTIONAL WHITE TABLE GRAPES

The effects of organic *versus* conventional production of table grapes grown at two locations were evaluated at harvest and during 14 days of cold storage at 0 °C (Amodio *et al.*, unpublished). Respiration rate, firmness, color, soluble solids content, acidity, organic acids, ascorbic acid, total phenols, antioxidant activity, flavonols, and peel characteristics were determined at harvest and during cold storage. Only from one location, phenolic compounds were significantly higher in the organic table grapes (505.1 ± 52.4 mg gallic acid/100 g) than in the conventional ones (370 ± 58 mg gallic acid/100 g). Similarly, antioxidant activity was higher in organic grapes (1211 ± 134 mg Trolox/100 g) than in conventional ones (763 ± 98 mg Trolox/100 g). This difference was observed until the sixth day of cold storage. Vitamin C in organic and conventional table grapes was similar at harvest and throughout cold storage. Storability was greatly affected by the agricultural system resulting in a shorter shelf-life for organic grapes, which scored the highest values for firmness and appearance.

6. ORGANIC *VS.* CONVENTIONAL PRODUCTS: COMPARATIVE ANALYSIS IN A "CUL DE SAC"

Organic and conventional production systems are comprised of a number of very diverse approaches: sequences of crop rotation, sources and quality of inputs, timing and doses of water and fertilizers, weed management and cover crop use, and pathogen and pest control. Organic and conventional farming cannot indeed be defined as two isolated clusters of agronomic practices. On one hand, it is possible to have high or moderate input intensity conventional farming systems. An example of the latter is the restricted use of chemical inputs by private standards (*i.e.*, for commercial reasons), such as for "zero pesticide residues" labels. In contrast, organic farming relies on more complex agro-ecosystem management that includes leguminous crops and organic matter-based amendments, although it may also be managed with a simple input substitution based on the replacement of chemical fertilizers and pesticides with organic-allowed inputs. In the scientific literature, this process in known as "conventionalization" of the organic farming [31] which may be summarized as the development of organic farming practices that might not be sustainable but that are not excluded by organic regulations [32]. In particular, organic greenhouse production is a clear example of this. The highly intensive production levels recorded in greenhouse systems requires high availability of nutrients to sustain crop growth. As a consequence, organic growers simplify crop rotation, excluding cover crops, and use easily soluble organic fertilizers. These issues should be taken into account in studies that aim to compare the quality and post-harvest performances of organic *vs.* conventional farming systems. Comparative analyses have to face the heterogeneity of the circumstances; observed differences and/or similarities would not necessarily be related to certification systems of the products because organically-certified products might have been produced by a "conventionalized" organic system and vice-versa. If organic-conventional system comparisons implemented the same agronomic practices and only varied in the type of nutrient or pesticide input, this would not be an organic *vs.* conventional system comparison but a comparison of organic *vs.* conventional inputs [33]. In this respect, there is room to begin comparative analyses among different approaches to organic methods of production in order to appreciate, within the greater heterogeneous cluster of organic practices, the impact on the quality of an organic conventionalized system *vs.* other organic systems based on agro-ecological practices. Such comparison studies should be considered as investigations of the effects of pre-harvest practices on post-harvest quality and product shelf life. In this regard, two research studies are reported below as preliminary results of this promising research approach.

7. EFFECTS OF DIFFERENT SYSTEMS OF ORGANIC PRODUCTION ON QUALITY AND POST-HARVEST OF TOMATO

Tomato, a climacteric fruit, represents a significant source of folate, vitamin C, polyphenols, and other antioxidants [34]. A conventionalized organic production system based on organic commercial fertilizers was compared with two organic production systems based on agro-ecological practices represented by (i) animal manure amendments and dead mulches of cover crops; or (ii) on-farm compost amended green manuring of cover crops. For the three production systems, tomato fruit respiration rate, morphological, physical, sensorial characteristics, and nutritional compound content were monitored. It was observed that dehydroascorbic acid was significantly higher in the conventionalized system for tomato fruit harvested at breaker and pink stages. However, this difference disappeared after 10 days of cold storage at 15 °C. Organic agro-ecological systems were able to obtain a similar tomato yield and fruit quality as the "conventionalized" organic system which used off-farm inputs only. This was confirmed both at harvest and during cold storage. Further information and results of this experiment have been reported [35]. It is worth noting that when the environmental conditions are similar in terms of climate, soil characteristics, availability of water, and soluble nutrients for the roots, any differences in the cultural practices (even relevant ones such as cover crop mixtures *vs.* bare soil, organic dead mulch *vs.* plastic mulch, manure *vs.*commercial organic fertilizers) did not affect crop quality and postharvest performance.

8. EFFECTS OF DIFFERENT SYSTEMS OF ORGANIC PRODUCTION ON QUALITY AND POST-HARVEST OF STRAWBERRY

Ceglie *et al.* [20] compared quality characteristics and postharvest performance of organically- and conventionally-produced strawberry fruit. Three organic farming systems (two "agro-ecological" and one "conventionalized") were compared to a conventional system. The production system affected the quality parameters of the strawberries cultivated in unheated tunnels. Conventionally-grown strawberry fruit were greater in diameter, chroma values, and firmness compared to the organically-produced fruit regardless of the specific organic system. Vitamin C, malic acid, tartaric acid, fructose, and glucose content were higher in the organic strawberries compared to conventional fruit. In fact, organically-grown strawberries received the highest evaluation for sweetness and aroma. Similar results were obtained for different varieties in the same geographic area (South of Italy) as reported by Conti *et al.* [21]. During the entire period of cold storage titratable acidity, citric acid, and total phenol values presented a similar decreasing trend from all of the production systems. However, the two organic agro-ecological systems differed in some aspects from both the "conventionalized" organic and the conventional system of production. At harvest, ascorbic acid and sucrose concentration were higher in both the organic agro-ecological than in the other two systems. Furthermore, during the entire cold storage period, tartaric acid and total phenols were higher in fruit from the organic agro-ecological systems than in either the conventional or the "conventionalized" organic systems. Thus, a "conventionalized" organic system may be halfway between conventional farming and organic-agro-ecological systems both in terms of quality and in terms of environmental sustainability. In this regard, Reganold *et al.* [36] linked the high quality of organic strawberry fruit with the high capability and stress resilience of organically-managed soil. Further studies on other commodities and with a longer-term assessment perspective are necessary to individuate sets of cultural practices applicable under the organic regulation which may enhance the quality of organic produce.

9. CONCLUSIONS

This review represents a small contribution to the wider picture of the quality of horticulture produce resulting from systems of production. Organic production systems have the objective of including a rational use of natural resources with high quality and shelf life performance. In many cases, conventional farming systems have achieved such high results. Nowadays, organic agriculture is increasing in terms of area of production and number of operators. A wide set of solutions have been proposed which, although valid with respect to organic certification standards, still need scientific assessment concerning claims of sustainability and high quality production. More in-depth analyses may relate the organic *vs.* conventional comparison to the more general issue of pre-harvest effects on postharvest performance of crops. In this respect, the balance between primary and secondary metabolic pathways seem to be an important aspect resulting from the complex interaction of genotype, environment, and agricultural practices which lead to differences in quality and postharvest performance of fresh fruits and vegetables. The need to improve the quality of food available in the world in a sustainable way should orient research on agricultural practices to increase the nutritional composition of fresh fruits and vegetables and to enhance the shelf life.

AUTHOR CONTRIBUTIONS

This work was a product of the combined effort of all of the authors. The authors equally contributed to the literature reviewing, to the manuscript writing and revisions.

CONFLICTS OF INTEREST

The authors declare no conflict of interest.

REFERENCES

1. Oates, L.; Cohen, M.; Braun, L.; Schembri, A.; Taskova, R. Reduction in urinary organophosphate pesticide metabolites in adults after a week-long organic diet. Environ. Res. 2014, 132, 105–111.

2. Rembiałkowska, E. Quality of plant products from organic agriculture. J. Sci. Food Agric. 2007, 87, 2757–2762.

3. Smith Spangler, C.; Brandeau, M.L.; Hunter, G.E.; Bavinger, J.C.; Pearson, M.; Eschbach, P.J.; Sundaram, V.; Liu, H.; Schirmer, P.; Stave, C.; et al. Are organic foods safer or healthier than conventional alternatives? A systematic review. Ann. Intern. Med. 2012, 157, 348–366.

4. Gill, S.S.; Tuteja, N. Reactive oxygen species and antioxidant machinery in abiotic stress tolerance in crop plants. Plant Physiol. Bioch. 2010, 48, 909–930.

5. Winter, C.K.; Davis, S.F. Organic foods. J. Food Sci. 2006, 71, R117–R124.

6. Woëse, D.; Lange, C.; Boess, K.; Bogl, W. A comparison of organically and conventionally grown foods—Results of a review of the relevant literature. J. Sci. Food Agric. 1997, 74, 281–293.

7. Lairon, D. Nutritional quality and safety of organic food. A review. Agron. Sustain. Dev. 2010, 30, 33–41.

8. Barański, M.; Średnicka-Tober, D.; Volakakis, N.; Seal, C.; Sanderson, R.; Stewart, G.B.; Leifert, C. Higher antioxidant and lower cadmium concentrations and lower incidence of pesticide residues in organically grown crops: A systematic literature review and meta-analyses. Br. J. Nutr. 2014, 112, 794–811.

9. Caruso, G.; Conti, S.; La Rocca, G. Influence of crop cycle and nitrogen fertilizer form on yield and nitrate content in different species of vegetables. Adv. Hortic. Sci. 2011, 25, 81–89.

10. Sikora, M.; Hallmann, E.; Rembiałkowska, E. The content of bioactive compounds in carrots from organic and conventional production in the context of health prevention. Rocz. Państw. Zakł. Hig. 2009,60, 217–220.

11. Del Amor, F.M.; Serrano-Martinez, A.; Fortea, I.; Nunez-Delicado, E. Differential effect of organic cultivation on the levels of phenolics, peroxidase and capsidiol in sweet peppers. J. Sci. Food Agric.2008, 88, 770–777.

12. Pieper, J.R.; Barrett, D.M. Effects of organic and conventional production systems on quality and nutritional parameters of processing tomatoes. J. Sci. Food Agric. 2009, 89, 177–194.

13. Rossi, F.; Godani, F.; Bertuzzi, T.; Trevisan, M.; Ferrari, F.; Gatti, S. Health-promoting substances and heavy metal content in tomatoes grown with different farming techniques. Eur. J. Nutr. 2008, 47, 266–272.

14. Krejčová, A.; Návesník, J.; Jičínská, J.; Černohorský, T. An elemental analysis of conventionally, organically and self-grown carrots. Food Chem. 2016, 192, 242–249.

15. Lundegårdh, B.; Mårtensson, A. Organically produced plant foods—Evidence of health benefits. Acta Agric. Scand. B 2003, 53, 3–15.

16. Luthria, D.; Singh, A.P.; Wilson, T.; Vorsa, N.; Banuelos, G.S.; Vinyard, B.T. Influence of conventional and organic agricultural practices on the phenolic content in eggplant pulp: Plant-to-plant variation.Food Chem. 2010, 121, 406–411.

17. Chassy, A.W.; Bui, L.; Renaud, E.N.C.; Van Horn, M.; Mitchell, A.E. Three-year comparison of the content of antioxidant micronutrients and several quality characteeristics in organic and conventionally managed tomatoes and bell peppers. J. Agric. Food Chem. 2006, 54, 8244–8252.

18. D'Evoli, L.; Tarozzi, A.; Hrelia, P.; Lucarini, M.; Cocchiola, M.; Gabrielli, P.; Franco, F.; Morroni, F.; Catelli-Forti, G.; Lombardi Boccia, G. Influence of cultivation system on bioactive molecules synthesis in strawberries: Spin-off on antioxidant and antiproliferative activity. J. Food Sci. 2010, 75, 94–99.

19. Bourn, D.; Prescott, J. A comparison of the nutritional value, sensory qualities, and food safety of organically and conventionally produced foods. Crit. Rev. Food Sci. Nutr. 2002, 42, 1–34.

20. Ceglie, F.G.; Mimiola, G.; Dechiara, M.L.; Amodio, M.L.; Colelli, G. A comparative study of quality and post-harvest performances of organically and conventionally grown strawberry (Fragaria × ananassa DUCH. "Festival"). In Proceedings of the 2nd Congress of the Italian Network for Research in Organic Farming—RIRAB, Rome, Italy, 11–13 June 2014.

21. Conti, S.; Villari, G.; Faugno, S.; Melchionna, G.; Somma, S.; Caruso, G. Effects of organic vs.conventional farming system on yield and quality of strawberry grown as an annual or biennial crop in southern Italy. Sci. Hort. 2014, 180, 63–71.

22. McGlone, V.A.; Jordan, R.B.; Seelye, R.; Clark, C.J. Dry-matter—A better predictor of the post-storage soluble solids in apples? Postharvest Biol. Technol. 2003, 28, 431–435.

23. DeEll, R.; Prange, K. Postharvest physiological disorders, diseases and mineral concentrationsof organically and conventionally grown McIntosh and Cortland apples. Can. J. Plant Sci. 1993, 73, 223–230.

24. Weibel, F.P.; Treutter, D.; Graf, U.; Haseli, A. Sensory and health related fruit quality of organic apples. A comparative field study over three years using conventional and holistic methods to assess fruit quality. In Proceedings of the 11th International Conference on Cultivation Technique and Ohytopatological Problems in Organic Fruit Growing, Weinsberg, Germany, 3–5 February 2004; pp. 185–195.

25. Nyanjage, M.O.; Wainwright, H.; Bishop, C.F.H.; Cullum, F.J. A comparative study on the ripening and mineral content of organically and conventionally grown Cavendish bananas. Biol. Agric. Hortic.2001, 18, 221–234.

26. Cayuela, A.; Vidueira, M.; Albi, A.; Gutierrez, F. Influence of the ecological cultivation of strawberries (Fragaria × ananassa cv. Chandler) on the quality of the fruit and on their capacity for conservation. J. Agric. Food Chem. 1997, 45, 1736.

27. Hasey, J.K.; Johnson, R.S.; Meyer, R.D.; Klonsky, K. An organic versus conventional farming system in kiwifruit. Acta Hort. 1997, 444, 223–228.

28. Benge, J.R.; Banks, N.H.; Tillmann, R.; Nihal de Silva, H. Pairwise comparison of the storage potential of kiwifruit from organic and conventional production system. N.Z. J. Crop Hort. Sci. 2000, 28, 147–152.

29. Uckoo, R.M.; Jayaprakasha, G.K.; Patil, B.S. Phytochemical analysis of organic and conventionally cultivated Meyer lemons (Citrus meyeri Tan.) during refrigerated storage. J. Food Compos. Anal. 2015,42, 63–70.

30. Amodio, M.L.; Colelli, G.; Hasey, J.K.; Kader, A.A. A comparative study of composition and postharvest performance of organically and conventionally grown kiwifruits. J. Sci. Food Agric. 2007,87, 1228–1236.

31. Goldberger, J.R. Conventionalization, civic engagement, and the sustainability of organic agriculture.J. Rural Stud. 2011, 27, 288–296.

32. De Wit, J.; Verhoog, H. Organic values and the conventionalization of organic agriculture. NJAS-Wagen. J. Life Sci. 2007, 54, 449–462.

33. Seufert, V.; Ramankutty, N.; Foley, J.A. Comparing the yields of organic and conventional agriculture.Nature 2012, 485, 229–232.

34. Charanjeet, K.; George, B.; Deepa, N.; Singh, B.; Kapoor, H. Antioxidant status of fresh and processed tomato. J. Food Sci. Technol. 2004, 41, 479–486.

35. Ceglie, F.G.; Muhadri, L.; Piazzolla, F.; Martinez-Hernandez, G.B.; Amodio, M.L.; Colelli, G. Quality and postharvest performance of organically grown tomato (Lycopersicon. Esculentum L. "Marmande") under unheated tunnel in Mediterranean climate. Acta Hort. 2015, 1079, 487–494.

36. Reganold, J.P.; Andrews, P.K.; Reeve, J.R.; Carpenter-Boggs, L.; Schadt, C.W.; Alldredge, J.R.; Ross, C.F.; Davies, N.M.; Zhou, J. Fruit and soil quality of organic and conventional strawberry agro-ecosystems. PLoS ONE 2010, 5, e12346.

6

EFFECT OF POSTHARVEST TREATMENTS ON QUALITY AND SHELF LIFE OF MANDARIN (CITRUS RETICULATA BLANCO)

Pradeep Raj Rokaya[1], Dilli Ram Baral[1], Durga Mani Gautam[1], Arjun Kumar Shrestha[1], Krishna Prasad Paudyal[2]

[1]Agriculture and Forestry University, Rampur, Chitwan, Nepal

[2]National Agricultural Research Institute (NARI), NARC, Khumaltar, Lalitpur, Nepal

ABSTRACT

Nepal produces large quantity of mandarin (Citrus reticulata Blanco) annually but faces big losses due to improper postharvest treatments. The experiment was carried out during 2012-2013 with the objective of assessing the effect of different postharvest treatments on quality and shelf life of mandarin. The postharvest treatments were: wax emulsion at 10%, bavistin at 0.1%, wax at 10% with bavistin at 0.1%, calcium chloride at 1%, Jeevatu at 5% along with untreated fruits as a control. The experiment was conducted for four weeks of storage at ambient condition in Nepal Agricultural Research Council, Khumaltar,

Kathmandu. Wax in combination with bavistin was found as the most effective in reducing the physiological loss in weight, whereas bavistin proved to reduce decay loss up to four weeks of storage. The fruits treated with wax alone and with combination of bavistin retained maximum firmness, juice percentage, vitamin C and palatability rating during the storage. The minimum total soluble solid and maximum titrable acidity were recorded in the fruits treated with wax plus bavistin. The present findings indicate that mandarin can be stored up to four weeks when treated with wax as well as wax with bavistin in the condition with 14°C - 18°C temperature and 45% - 73% relative humidity.

Keywords: Mandarin, Fruit Quality, Postharvest Treatments, Palatability Rating and Postharvest Loss in Weight

1. INTRODUCTION

Mandarin (Citrus reticulata Blanco) is a most promising fruit crop that stands in first position of the total fruit industry in Nepal. The country produces a large quantity of mandarin in its hilly areas from east to west comprising fifty four districts. It shares 0.97 percent in AGDP and 0.33 percent in GDP [1] . The country exports mandarin to India, China, Bangladesh, Bhutan, Pakistan and other countries about 600 tonnes annually [2] .

Since mandarin is a non-climacteric and perishable fruits, it cannot be kept for a long time during transportation and storage. Mandarin fruits can be kept hardly for 1 - 2 weeks depending upon temperature and humidity. The large volume of the losses starts right from harvesting and loss increases many folds during the postharvest steps. Worldwide postharvest loss in fruits and vegetables are as high as 30% - 40% and even much higher in developing countries like Nepal [3] . Improper postharvest handling practices lead to inferior fruit quality and fetch poor market price. In mandarin, loss of 20% - 25% has been reported during transportation from field to market [4] . The methods of harvesting, injury to fruit during harvesting, and weather conditions during harvest greatly determine the extent of decay losses during subsequent handling and storage [5] . Losses of mandarin fruits during the harvesting, handling and marketing are 25% - 30% [6] . Generally, mandarin fruits after harvest show the great losses and become non-accepted by consumer in the market. Loss is greatly influenced by the improper methods of harvesting, handling, packaging and transportation. Qualitative losses (such as loss of caloric and nutritive value, loss of acceptability by consumers, and loss of edibility) are more difficult to measure than quantitative losses of fresh fruits [7] . The prevalence of high inoculums of pathogens in the orchards due to poor plant protection measures besides non-adoption of scientific harvesting, handling, transit, storage, and antimicrobial treatments leads to heavy losses of fruits after harvest [8] . Various viable technologies like use of fungicides, cold storage, controlled atmosphere storage, anti-transpirants, wax coating, growth retardants, irradiation, and different types of packaging materials, etc. have been used to increase the shelf life of harvested fruits in past decades [9] . Postharvest treatments play a significant role in extending shelf life of the fruits [10] . Nepal faces big losses of mandarin annually due to not adopting proper postharvest handling practices during harvesting, transportation and storage. Hence, to minimise these postharvest losses and to maximise the quantitative and qualitative parameters along with prolongation of storage capacity, postharvest treatments with wax and other safe fungicides are urgent for effective marketing of mandarin in the country.

2. MATERIALS AND METHODS

The present study was carried out at Nepal Agriculture Research Council (NARC), Kathmandu to assess the different postharvest treatments during 2012-2013 with the objective of managing the postharvest losses during storage condition. Fully ripe, greenish yellow, well matured and healthy mandarin fruits were clipped and brought from Lamjung district and fruits were washed with tap water then air-dried under the shade. These cleaned fruits were graded according to uniformity in size and immersed in emulsion for two minute then dried under shade. Eighty fruits from each treatment were kept on tray and stored in ambient condition (16°C + 2°C and 45% - 73% RH). The experiment was laid out in completely randomised design comprising six treatments i.e. T_1 = untreated fruits (control), T_2 = wax 10%, T_3 = bavistin 0.1%, T_4 = wax 10% with bavistin 0.1%, T_5 = calcium chloride 1% and T_6 = jeevatu 5% (package of beneficial microbes) with four replications. The observations on various parameters were recorded at weekly intervals up to four weeks of storage. Respiratory losses in terms of fruit weight and decay loss were determined and calculated on the basis of percent weight loss and percent decayed fruits. Fruit firmness was taken with penetrometer (effigy oil model having 8 mm tip). As a quality parameter, TSS was determined with help of refractometer corrected at 20°C, acidity, and vitamin C was determined as per outlined by AOAC [11] . The palatability rating of the fruits was conducted by a panel of five judges on the basis of fruit colour, texture, aroma and taste using hedonic rating 1 - 5. Fruits scoring a minimum of 3 out of 5 were regarded as acceptable form from the consumer point of view. Finally, the data were tabulated in excel sheet and analysed with statistical method mentioned by Gomez and Gomez [12] .

3. RESULTS AND DISCUSSION

3.1. PHYSIOLOGICAL LOSS IN WEIGHT (PLW)

Physiological loss in weight (PLW) was significantly increased in all the treatments with the advancement of the storage period and the increasing trends in the weight loss percentage was found maximum in the fruits with untreated as a control in each week up to the 4th week of storage period. Minimum percentage of PLW was observed in the fruits treated with wax in combination with bavistin in all the weeks during storage and the losses ranged from 3.02% in the 1st week to 9.81% in the 4th week which was statistically at par with the findings of wax 10% (3.09 to 10.18%) whereas maximum weight loss was recorded in the fruits with control (6.41% to 21.41%) during the storage (Table 1).

This minimum weight loss in the wax treated fruits might be due to retardation in the process of transpiration and respiration by closing of lenticels and stomata of the cell wall of the fruits. The losses in fruit weight and moisture content of the peel were mainly caused by fruit transpiration in which water moved out and resulted in wilted rind and a shrivelled appearance [13] . Thus, wax emulsion might have been an effective treatment to reduce weight loss by checking the stomata and lenticels of the cell wall of the fruits which reduces the rate of transpiration and respiration. [14] claimed that wax coated fruit retained better glossiness and fresh appearance being a moisture barrier. These findings were in consonance with the report of the Bhusal [15] , Bastakoti and Gautam [16] in mandarin, Chaudhary and Dhaka [17] in kinnow mandarin, Deka et al. [10] in Khasi mandarin, Ahmad et al. [18] in kinnow mandarin, Yadav et al. [9] , in kinnow mandarin, and Bhullar [19] in sweet orange who found minimum weight loss in the fruits treated with wax emulsion.

3.2. Decay Loss

The perusal from data shown in Table 1 the decay loss increased significantly with the prolongation of storage period irrespective of the treatments. The decay loss was noticed from the first week of the storage in untreated fruits (control) and it was appeared from the second week in other treatments except bavistin. Decay loss was found significantly in the third week of the storage in all treatments. The minimum decay loss was noticed in the fruits with bavistin alone (0.7%) followed by combination of wax 10% + bavistin 0.1% (1.3%), and wax at 10% (3.6%) whereas the maximum decay loss was observed in the fruits with untreated (control) at the end of storage. The minimum decay loss in the bavistin and wax treated fruits might be due to inhibition of the moisture and microbial agent particularly. Yadav et al. [9] reported that the minimum decay loss might be due to interaction effect of wax coating as a microbial inhibitor as well as moisture inhibitor. Similar results on the decay loss were observed by Chaudhary and Dhaka [17] in kinnow mandarin, Mahajan et al. [14] in Asian Pear, Gautam et al. [20] in mango, Ahmad et al. [18] in kinnow mandarin, Bhusal [15] in mandarin, Hasan et al. [21] , Deka et al. [10] in Khasi mandarin, and Yadav et al. [9] in kinnow mandarin, who reported minimum decay loss using the wax and wax like safe fungicide during the storage.

3.3. Fruit Firmness

The fruit firmness was decreased with the advancement of the storage period in all the treatments.Table 2 shows that the decreasing trend was started from the first week to the end of storage in all the treatments. At the first week, the wax at 10% in combination with bavistin at 0.1% treated fruits resulted in the more intact (4.16 kg/cm^2) and the lowest was recorded in the fruits with control (3.78 kg/cm^2). At the end of the storage, the firmness was observed maximum in the fruits treated with wax 10% plus bavistin 0.1% (3.08 kg/cm^2) followed by wax 10% (3.03 kg/cm^2) and the lowest firmness was found in the fruits with control (2.09 kg/cm^2). The decline

Table 1. Effect of postharvest treatments on PLW and decay loss percentage.

Treatments	Physiological loss in weight (%) Weeks of storage				Decay loss (%) Weeks of storage			
	1	2	3	4	1	2	3	4
Control	6.41	12.86	17.44	21.41	1.89	4.46	7.39	16.70
Wax 10%	3.09	6.35	8.06	10.18	0.00	0.92	1.76	3.60
Bavistin 0.1%	5.50	10.9	14.29	17.18	0.00	0.00	0.27	0.70
Wax 10% + Bavistin 0.1%	3.02	5.98	7.84	9.81	0.00	0.00	0.56	1.30
$CaCl_2$ 1%	5.61	10.65	14.93	18.07	0.00	2.39	3.25	5.30
Jeevatu 5%	5.78	11.19	14.84	17.87	0.00	2.37	3.77	6.10
CD at 5%	0.85	1.14	1.19	1.47	NS	NS	4.14	7.40

NS: Non-significant.

Table 2. Effect of postharvest treatments on fruit firmness and juice recovery percentage.

Treatments	Fruit firmness (kg/cm^2) Weeks of storage				Juice recovery (%) Weeks of storage			
	1	2	3	4	1	2	3	4
Control	3.78	3.16	2.59	2.09	47.26	42.89	38.80	34.65
Wax 10%	4.07	3.17	3.37	3.03	49.49	47.64	45.87	43.45
Bavistin 0.1%	3.94	3.56	3.16	2.83	47.81	45.86	43.60	40.92
Wax 10% + Bavistin 0.1%	4.16	3.79	3.42	3.08	49.56	47.84	45.95	43.81
$CaCl_2$ 1%	3.91	3.53	3.12	2.68	47.57	45.30	42.44	39.78
Jeevatu 5%	3.88	3.49	3.03	2.61	48.09	45.71	42.83	40.66
CD at 5%	NS	0.35	0.45	0.40	NS	NS	4.19	4.62

NS: Non-significant.

in the firmness might be due to moisture loss from the fruits cells. The wax alone or in combination with bavistin showed the more intact (firmness) than other treatments. It might be due to wax alone or in combination with other fungicides acts as moisture and microbial inhibitor that reduces the respiration and transpiration of the fruits. Mahajan et al. [14] reported that wax coat fruits reduced the moisture loss from the surface, thus maintain cell wall integrity and tissue rigidity. Sidhu et al. [22] and Yadav et al. [9] reported that the firmness decreased with the increase in storage period.

3.4. Juice Recovery

Table 2 shows that the juice recovery percentage was decreased with time during the storage in all the treatments. Wax treated fruits recorded the maximum juice recovery percentage in all the weeks. In the first week, the higher juice percentage was obtained in the wax 10% plus bavistin 0.1% (49.56%) which was statistically at par with wax 10% (49.49%) as against control (47.26%). The trend in decrease in juice percentage during the storage was might be due to loss of moisture from the surface of the fruits. The minimum decrease in juice percentage was observed in the fruits treated with wax 10% plus bavistin 0.1% from the 1st week (49.56%) to the 4th week (43.81%) followed by wax 10% from the 1st week (49.49%) to the 4th week (43.45%) as against control from the 1st week (47.26%) to the 4th week (34.65%). There was significant decrease in the juice percentage of untreated fruits (control) than treated fruits. The fruits treated with wax and in combination with bavistin showed low reduction in juice content during storage as compared to other chemically treated fruits and control. This might be due the fact that the wax acted as a barrier which had checked the losses of the moisture from the fruit surface. Similar results were also reported by Yadav et al. [9] in kinnow mandarin, Sonkar et al. [8] in kinnow mandarin, Sharma and Ghuman [23] in kinnow fruits, and Deka et al. [10] in khasi mandarin.

3.5. Total Soluble Solid (TSS)

TSS is one of the major indicators to judge the quality of the mandarin fruits. As shown in Table 3, TSS increased with the increasing period of storage in all the treatments and the increasing trend was higher in untreated fruits (control) than the treated fruits with different chemicals. Untreated (control) fruits showed the maximum TSS content during the storage and ranged from the 1st week (10.92° Brix) to the 4th week (12.88° Brix) and minimum TSS was recorded in the fruits treated with wax 10% from the 1st week (10.35° Brix) to the 4th week (11.51° Brix) which was at par with wax 10% in combination with bavistin 0.1% from the 1st week (10.39° Brix) to the 4th week (11.65° Brix). The trend showed that wax treated fruits was significantly superior because of the gradual increment in the TSS change whereas in control, it was increased at faster pace.

The faster TSS increment in the untreated fruits might be due to faster metabolic activities through respiration and transpiration than in treated fruits with different chemicals. Jholgiker and Reddy [24] claimed that the gradual increment in TSS of fruits treated with coating material may be justified by the twin role of coating material, acting as a physical barrier for transpiration losses and creating a modified atmosphere resulting in building of internal CO_2 and depletion of O_2. Similar results were noticed by Sidhu et al. [22] in pear; Deka et al. [10] , Ladaniya et al. [25] in Nagpur mandarin; and Mahajan et al. [14] in pear.

Table 3. Effect of postharvest treatments on TSS and TA.

Treatments	TSS (°Brix)				Titrable acidity (%)			
	Weeks of storage				Weeks of storage			
	1	2	3	4	1	2	3	4
Control	10.92	11.45	12.10	12.88	0.86	0.75	0.65	0.53
Wax 10%	10.35	10.67	11.09	11.51	0.91	0.83	0.75	0.69
Bavistin 0.1%	10.44	11.06	11.30	11.87	0.90	0.82	0.75	0.67
Wax 10% + Bavistin 0.1%	10.39	10.79	11.44	11.65	0.93	0.86	0.78	0.73
$CaCl_2$ 1%	10.35	10.99	11.36	11.82	0.89	0.83	0.74	0.67
Jeevatu 5%	10.62	11.06	11.60	11.99	0.88	0.79	0.72	0.65
CD at 5%	NS	NS	NS	0.60	NS	0.70	0.07	0.06

NS: Non-significant.

3.6. Titrable Acidity (TA)

The data presented in Table 3 revealed that the results pertaining to the effect of different levels of treatments to the titrable acidity (TA) was significant at the end of the storage of mandarin. The TA significantly decreased with the advancement of the storage period. The decreasing trend of acidity during the storage period was probably due to utilization of acid in tricarboxylic acid cycle in respiration process. The TA was recorded maximum in the fruits treated with wax 10% plus bavistin 0.1% (0.73%) at par with wax 10% (0.69%) as against control (0.53%) at the end of the storage. The higher acidity in the wax treated fruits might be due to lesser utilization of the acids in the respiration process during the storage whereas untreated fruits had minimum acids was might be due to faster utilization of the acids in the respiration process during storage. The results are in line with the findings reported by Sonkar et al. [8] in kinnow mandarin, Jholgiker and Reddy [24] in Annona, Sidhu et al. [22] in pear, and Deka et al. [10] in khasi mandarin.

3.7. Vitamin C

A perusal from the data presented in Table 4 pertaining the effect of treatments to the vitamin C was found significant and decreasing with the advancement of the storage period in all the treatments. The maximum vitamin C was recorded in the fruits treated with wax 10% in combination with 0.1% bavistin (33.26 mg/100ml) which is statistically at par with wax 10% (32.85 mg/100ml) as against control (29.28 mg/100ml) in the first week. Likewise, vitamin C wax maximum in the fruits treated with wax 10% plus bavistin 0.1% (24.65 mg/100ml) which was statistically at par with wax 10% (24.16 mg/100ml) as against of control (18.86 mg/100ml). The decreasing trend in the vitamin C was probably due to degradation of the ascorbic acid during the storage. Lee and Kader [26] suggested that loss of vitamin C is caused by leaching in surrounding water and thermal breakdown. The retention of the vitamin C in the wax treated fruits might be due to less degradation of the ascorbic acid in the storage. These findings were supported with findings of Sonkar et al. [8] in kinnow mandarin, Deka et al. [10] in Khasi mandarin, Ladaniya et al. [25] in Nagpur mandarin, Ahmad et al. [18] in kinnow mandarin who reported that wax treated fruits retained the maximum vitamin C content.

3.8. Palatability Rating

The fruits treated with different chemicals were evaluated in respect of palatability and the rating was found decreasing trend during the storage. Table 4 shows that Mandarin flavour quality is often rapidly lost after fruits are picked or commercially packed. The highest rating (4.56) was recorded in the fruits treated with wax 10% alone and in combination with bavistin which was statistically at par with the other chemically treated fruits as against of control (4.22) in the 1st week. However, rating was found maximum in the fruits treated with wax 10% plus bavistin 0.1% (3.48) which was statistically at par with wax 10% (3.46) as against of control (2.43). The treated fruits had showed the good appearance, sweet taste and desirable flavour and acceptable position during the storage. Among the treatments, only wax 10% plus bavistin 0.1% (3.48) followed by wax 10% (3.46) had showed the acceptable position. The lowest rating was recorded in the fruits with control (2.43) which is below of the acceptable position. The decrease in the palatability with time was might be due to ascribed to certain bio-chemical changes in the fruits.

Table 4. Effect of postharvest treatments on vitamin C and sensory evaluation.

Treatments	Vitamin C (mg/100ml)				Sensory evaluation (1 - 5 scale)			
	Weeks of storage				Weeks of storage			
	1	2	3	4	1	2	3	4
Control	29.28	25.37	22.22	18.86	4.22	3.68	3.06	2.43
Wax 10%	32.85	29.12	25.89	24.16	4.56	4.25	3.93	3.46
Bavistin 0.1%	31.42	27.99	24.63	22.54	4.40	4.09	3.64	3.22
Wax 10% + Bavistin 0.1%	33.26	29.62	26.46	24.65	4.56	4.26	3.93	3.48
$CaCl_2$ 1%	32.61	28.23	24.84	22.67	4.32	4.02	3.57	3.15
Jeevatu 5%	30.34	26.45	23.40	21.10	4.28	4.01	3.53	3.12
CD at 5%	NS	NS	NS	2.91	0.15	0.17	0.24	0.30

NS: Non-significant

Chemically treated fruits had acceptable colour, glossy appearance, sweet taste, and good flavour that might be due to the fact that chemical treatments acts a barrier to the moisture loss and rapid respiration and transpiration loss. Control fruits had higher rate of metabolic activities with respiration and transpiration, thus dislike by the people. Orange-like flavour decreased not as a consequence of an increment in the ethanol content, but because of other, minimal chemical changes that took place during storage [27] . From the above facts it can be concluded that waxed fruits had fresh, glossy external which is highly acceptable in the market up to the 4th week of storage and fruits without treatments observed as dull and shrivelled appearance and those could not acceptable up to the 2nd week. The findings were in corroboration with the findings of Sindhu et al. [22] in pear, Choudhary and Dhaka [17] in kinnow mandarin, Mahajan et al. [14] in pear, Karibasappa and Gupta [28] in khasi mandarin, Singhrot et al. [29] in lemon who reported that palatability was decreased with time during the storage.

4. CONCLUSION

From findings drawn as above, we can conclude that postharvest losses can be minimised by treating mandarin fruit with different fungicides to prolong the shelf life and to enhance fruit quality. Wax alone and in combination with bavistin reduced the weight and decay loss as well as improved the internal quality of the fruits. Fruits treated with wax and with bavistin can be stored up to four weeks at the condition of 14°C - 18°C temperature and 45% - 73% RH. This study will be a milestone to use different postharvest treatments to reduce postharvest losses in the storage.

ACKNOWLEDGEMENTS

The authors are thankful to Director General of Department of Food Technology and Quality Control (DFTQC) of Ministry of Agricultural Development, and to the chief of postharvest research unit of NARC, Khumaltar, Lalitpur for providing laboratory facility during the experiment.

REFERENCES

1. PMD (2012) Postharvest Management and Value Addition of Fruits in Production Catchments in Nepal. A Project Submitted to SAARC Development Fund (SDF) Secretariat, Postharvest Management Directorate, Shreemahal, Kathmandu, Nepal, 1-16.

2. TEPC (2013) Statistical Information. Trade and Export Promotion Center, Lalitpur, Kathmandu, Nepal.

3. Bhattarai, R.M., Rijal, R.K. and Mishra, P. (2013) Postharvest Losses in Mandarin Orange: A Case Study of Dhankuta District of Nepal. African Journal of Agricultural Research, 8, 763-767.

4. PHLRD (2005) Food Loss Status in Nepal. Progress Report. Postharvest Loss Reduction Division, Postharvest Management Directorate, Kathmandu, Nepal, 3.

5. Ladaniya, M.S. (2008) Citrus Fruit: Biology, Technology and Evaluation. Elsevier Inc., Atlanta, USA, 1-10.

6. Singh, S. (2001) Citrus Industry of India. In: Singh, S. and Naqvi, S.A.M.H., Eds., Citrus, International Book Distributing Company, Lucknow, India, 1-44.

7. Kader, A.A. (2005) Increasing Food Availability and Reducing Postharvest Losses of Fresh Produce. Proceedings of the 5th International Postharvest Symposium, Verona, 6-11 June 2004, 2169-2175.

8. Sonkar, R.K., Sarnaik, D.A., Dikshit, S.N. and Saxena, R.R. (2009) Individual Stretch Cling Film Wrapped Kinnow Mandarin under Ambient Storage. Indian Journal of Horticulture, 66, 22-27.

9. Yadav, M., Kumar, N., Singh, D.B. and Singh, G.K. (2010) Effect of Postharvest Treatments on Shelf Life and Quality of Kinnow Mandarin. Indian Journal of Horticulture, 67, 243-248. [Paper reference:6]

10. Deka, B.C., Sharma, S. and Borah, S.C. (2006) Postharvest Management Practices for Shelf Life Extension of Khasi

Mandarin. Indian Journal of Horticulture, 63, 251-255. [Paper reference:7]

11. A.O.A.C. (2005) Official Methods of Analysis. Association of Official Analytical Chemists, Washington DC, USA.

12. Gomez, A.K. and Gomez, A.A. (1984) Statistical Procedures for Agricultural Research. John Wiley and Sons Inc., New York.

13. Wills, R., McGlasson, B., Graham, D. and Joyce, D. (2007) Postharvest: An Introduction to the Physiology and Handling of Fruit, Vegetables and Ornamentals. 2nd Edition, University of New South Wales Press, Sydney.

14. Mahajan, B.V.C., Datta, A.S. and Dhillon, W.S. (2004) Effect of Pre-Storage Treatments on the Quality and Storage Life of Asian Pear. Indian Journal of Horticulture, 61, 342-344. [Paper reference:5]

15. Bhusal, Y. (2002) Improvement of Postharvest Shelf Life of Mandarin Orange (Citrus reticulate Blanco), M.Sc. Thesis, Institute of Agriculture and Animal Science (IAAS), Rampur, Chitwan, Nepal.

16. Bastakoti, P. and Gautam, D.M. (2007) Effect of Maturity Stages and Postharvest Stages on Shelf Life and Quality of Mandarin Orange in Modified Cellar Store. Journal of Institute of Agriculture and Animal Science, 28, 65-74.

17. Choudhary, M.R. and Dhaka, R.S. (2005) Effect of Different Postharvest Treatments on Quality of Kinnow Mandarin Fruits during Storage. Haryana Journal of horticulture Science, 34, 39-41.

18. Ahmad, M.S., Thakur, K.S. and Kaushal, B.B.L. (2005) Post-Harvest Treatments to Reduce Postharvest Losses in Kinnow Mandarin. Indian Journal of Horticulture, 62, 63-67.

19. Bhullar, J.S. (1981) Effect of Wax Emulsion, Benlate and Some Growth Regulators on the Storage Life of Pineapple Sweet Orange. Haryana Journal Horticulture Science, 10, 147-150.

20. Gautam, B., Sarkar, S.K. and Reddy Y.N. (2003) Effect of Post-Harvest Treatments on Shelf Life and Quality of Banganapalli Mango. India Journal of Horticulture, 60, 135-139.

21. Hassan, Z.H., Lesmayati, S., Qomariah, R. and Hasbianto, A. (2014) Effects of Wax Coating Applications and Storage Temperatures on the Quality of Tangerine Citrus (Citrus reticulata) var. Siam Banjar. International Food Research Journal, 21, 641-648.

22. Sidhu, G.S., Dhillon, W.S. and Mahajan, B.V.C. (2006) Effect of Waxing and Packaging on Storage and Shelf Life of Pear cv. Punjab Beauty. Haryana Journal of Horticulture Science, 35, 51-56.

23. Sharma, A. and Ghuman, B.S. (2009) Evaluation of On-Farm Waxing and Grading Prototype for Fresh Kinnow Fruits. Indian Journal of Horticulture, 66, 288-290.

24. Jholgiker, P. and Reddy, B.S. (2007) Effect of Different Surface Coating Material on Post-Harvest Physiology of Annona squamosa L. Fruits under Ambient and Zero Energy Cool Chamber Storage. Indian Journal of Horticulture, 64, 41-44.

25. Ladaniya, M.S., Singh, S. and Mahalle, B. (2005) Sub-Optimum Low Temperature of "Nagpur" Mandarin as Influenced by Wax Coating and Intermitant Waxing. Indian Journal of Horticulture, 62, 1-7.

26. Lee S.K. and Kader, A.A. (2000) Pre-Harvest and Postharvest Factors Influencing Vitamin C Content of Horticultural Crops. Postharvest Biology and Technology, 20, 207-220. http://www.elsevier.com/locate/postharvbio

27. Marcilla, A., Zarzo, M. and del Río, M.A. (2006) Effect of Storage Temperature on the Flavour of Citrus Fruit. Spanish Journal of Agricultural Research, 4, 336-344.

28. Karibasappa, G.S. and Gupta, P.N. (1988) Storage Studies of Khasi Mandarins. Haryana Journal of Horticulture Science, 17, 196-200.

29. Singhrot, R.S., Sharma, R.K., Sandooja, J.K. and Singh, J.P. (1987) Effect of Some Chemicals to Enhance Shelf Life of Baramasi lemon. Haryana Journal of Horticulture Science, 16, 25-30.

7

DEGRADATION AND DETOXIFICATION OF BANANA POSTHARVEST TREATMENT WATER USING ADVANCED OXIDATION TECHNIQUES

Dunia E. Santiago, Elisenda Pulido Melián, Cristina Fernández Rodríguez, José A. Ortega Méndez, Sebastián O. Pérez-Báez, José M. Doña-Rodríguez*

Grupo de Photocatálisis y Espectroscopía Aplicada al Medioambiente-FEAM (Unidad Asociada al Instituto de Ciencia de Materiales de Sevilla, C.S.I.C.), CIDIA-Depto. de Química, Edificio del Parque Científico Tecnológico, Universidad De Las Palmas De Gran Canaria, Las Palmas, Spain

ABSTRACT

The elimination and detoxification of water contaminated with the principal commercial fungicides—thiabendazole (Textar 60 T) and imazalil sulphate (Fruitgard IS 7.5)—used in the postharvest treatment of bananas was successfully achieved through the use of advanced oxidation processes (AOPs). The techniques tested were heterogeneous photocatalysis with TiO_2 and TiO_2-activated carbon (TiO_2-CA) and Fenton and photo-Fenton processes. The evolution of fungicide concentration during degradation was studied together with mineralization and toxicity levels. When treated with any of the methods under study, the water meets the conditions stipulated by the legislation currently in force for its reuse in irrigation. However, this study proposes that the Fenton process is the best option for treatment of this type of water due to the efficiency of its reactions and its low economic cost.

Keywords: Thiabendazole, Imazalil Sulphate, Photocatalysis, Fenton, Photo-Fenton, Banana Postharvest

1. INTRODUCTION

The water under study represents a typical wastewater obtained after bananas have been washed as part of their postharvest treatment. In this postharvest treatment the fruit are normally washed two or three times in order to remove larger particles and any dust that might have been brought with the fruit from its growing area. A fungicide mixture is also applied to combat possible diseases, particularly crown rot, that might affect the fruit. Among the most commonly used fungicide mixtures are those studied in this work which are thiabendazole-based or imazalil sulphate-based [1].

This water contaminated with the postharvest fungicide is normally washed away into the sewage system. For a banana packing system of between 10 and 60 tons/ day, this represents a total daily volume of some 10000 L of water that needs to be washed away. The various wastewater samples analysed for this study contained concentrations of between 10 and 15 $mg \cdot L^{-1}$ of imazalil sulphate or between 25 and 30 $mg \cdot L^{-1}$ of thiabendazole. By law, this fungicide concentration must be removed before it is washed away into the sewer system. Legislation in the Canary Islands (Spain), one of the biggest banana producing regions, requires that concentrations of this type of substance must not exceed 0.05 $mg \cdot L^{-1}$ [2]. However, because of the growing scarcity of the water resource in the Canary Islands as a result of overexploitation and contamination of the aquifers, the use of treated water for irrigation purposes is of particular interest. For this to be possible, restrictions related to quality indicators such as BOD_5, COD, nitrates, etc., must also be complied with in addition to the required fungicide threshold concentrations [3].

Studies have been published on the effectiveness of advanced oxidation techniques, specifically heterogeneous TiO_2-based photocatalysis and the Fenton processes for the degradation of pesticides and fungicides in water [4-6]. These processes, although making use of different reacting systems, are all characterised by the same chemical feature: production of ·OH radicals [7].

In heterogeneous photocatalysis radiant energy is absorbed by the heterogeneous photocatalyst, a broadband semiconductor, in this case TiO_2. The degradation reactions of the pollutants take place in the interfacial region without the catalyst undergoing chemical changes.

To improve the efficiency of the TiO_2, an inert surface like activated carbon is used as a support for the TiO_2 to provide a large surface area. There are two main reasons for using adsorbents like this as support for the photoactive catalyst: firstly, to increase the size of the photoactive surface and secondly, to promote contact between the compound to be degraded and the catalyst by preconcentration of the pollutant at the semiconductor surface [8].

Studies have been undertaken which report that activated carbon increases not only the accessible surface area of the TiO_2, as it impedes its aggregation, but also its decantability, acid-base properties and absorbance spectrum. This all leads to an improved photocatalytic performance. It has been established that the amount of activated carbon that maximises these benefits lies between 7% and 13% [9].

Fenton's reagent is an aqueous solution of hydrogen peroxide and ferrous ions which provides an important source of hydroxyl radicals and which, under acidic conditions (pH 2 - 4), is a powerful oxidizing agent of organic compounds. Degradation rates increase substantially when the process is complemented with UV/visible radiation. In these cases it is known as a photo-Fenton process. The process becomes a photocatalytic one, as the Fe^{2+} oxidized to Fe^{3+} in the Fenton reaction is reduced again to Fe^{2+} by the action of radiation. Both reactions generate hydroxyl radicals [10].

The Fenton process has proven to be effective for the degradation of various compounds, including chlorinated aliphatic and aromatic compounds, nitroaromatics, azodyes, chlorobenzene, phenols, formaldehyde, herbicides, soil pollutants, etc. [11]. Fenton's reagent has also been successfully applied in the reduction of COD in municipal waste and ground waters and in leachate treatment. It is useful for the pre-treatment of non-biodegradable compounds [12]. Among the very few compounds that cannot be attacked by this reagent are acetone, acetic acid, oxalic acid, paraffins and organochlorine compounds [13].

The use of Fenton and photo-Fenton processes in the treatment of pesticide-containing wastewater has also been studied: the degradation of triazine herbicides [14], methyl-parathion [15], fenuron [16], the fungicides pyrimethanil and triadimenol [17], and commercial mixtures of pesticides [6, 18] are known examples.

The objective of this study is the removal of both thiabendazole and imazalil sulphate from the water used in the postharvest washing of bananas, using advanced oxidation techniques to comply with legislation governing wastewater disposal and/or water reuse for agricultural purposes. The effectiveness of four techniques is compared: heterogeneous photocatalysis based on TiO_2 used alone and in combination with activated carbon, Fenton and photo-Fenton processes.

2. EXPERIMENTAL

2.1. Characterisation of the Water to be Treated

Of the two fungicide mixtures under study in this work, one is thiabendazole-based ($C_{10}H_7N_3S$) and the other imazalil sulphate-based ($C_{14}H_{16}Cl_2N_2O_5S$), shown in **Figure 1**. The latter belongs to the imidazole chemical group and was developed to control postharvest diseases in fruit, particularly when there are problems of pathogen tolerance to benzimidazole fungicides like thiabendazole [19].

The fungicide mixtures were provided by the North Gran Canaria Agricultural Cooperative (Cooperativa Agrí- cola del Norte de Gran Canaria), which forms part of the Canary Island Banana Cooperative (COPLACA). The thiabendazole-based fungicide mixture was a 0.075%

(v/v) solution of Textar 60 T which contained 60% (w/v) of thiabendazole. The imazalil sulphate based fungicide mixture was a 0.6% (v/v) solution of Fruitgard IS 7.5. The wastewater samples analysed in this study revealed concentrations of 25 - 30 $mg \cdot L^{-1}$ of thiabendazole and 10 - 15 $mg \cdot L^{-1}$ of imazalil sulphate. Other chemical characteristics of both waste products are shown in **Table 1**.

2.2. Materials and Methods

2.2.1. Reagents/Chemicals

The reagents used for the photocatalytic tests were: TiO_2 (Degussa P25) and activated carbon supplied by Panreac (surface area: 778.28 m^2/g, pore volume: 0.454 cm^3/g, ash content: <1%, impurities: $Cl^- \approx 0.2\%$).

35% H_2O_2 (Scharlau) was used for the Fenton tests. Panreac dehydrated ferrous sulphate (99%) was used as the Fe^{2+} source.

pH was adjusted with diluted H_2SO_4 and NaOH.

Catalase was used to remove the H_2O_2 and avoid interferences with the analytical results (Micrococcus Lysodeikticus obtained from Fluka with 243075 $U \cdot mL^{-1}$).

2.2.2. Analytical Determinations

Remaining concentrations of thiabendazole at different reaction times were HPLC-measured using a Supelco Discovery C18 (25 cm × 4.6 mm ID, 5 μm particles) and an acetonitrile-water-ammonia 40:60:0.6 solution as mobile phase, using a UV detector (λ = 280 nm). Concentrations of imazalil sulphate were measured by the same method, using in this case an acetonitrile-10mM KH_2PO_4solution (35:65) and 200 $mg \cdot L^{-1}$ of sodium 1-octanedulfonate as mobile phase (adjusted to pH 3 with phosphoric acid), with detection conducted at a wavelength of λ = 225 nm. The quantification of both fungicides was performed by least-squares fit [20].

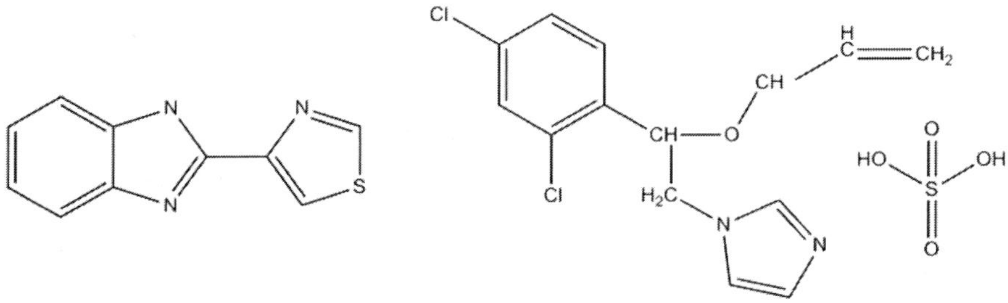

Figure 1. Thiabendazole molecule (left) and imazalil sulphate molecule (right).

Table 1. Parameters of interest of the water under study. Thiabendazole-based or imazalil-based fungicide mixtures.

Parameters	Thiabendazole	Imazalil sulphate
Concentration ($mg \cdot L^{-1}$)	25 - 30	10 - 15
pH	6.32	5.22
TOC ($mg \cdot L^{-1}$)	11 - 14	11 - 12
Toxicity (% Inh.)	≈77%	≈75%
BOD_5 ($mg \cdot L^{-1}$)	20	20
COD ($mg \cdot L^{-1}$)	557.83	580.64
SS ($mg \cdot L^{-1}$)	128	120
Nitrates ($mg \cdot L^{-1}$)	0.01	0.01
Nitrites ($mg \cdot L^{-1}$)	0.00	0.00
Sulphates ($mg \cdot L^{-1}$)	4.08	4.08
Chlorides ($mg \cdot L^{-1}$)	2.52	2.52

Total organic carbon (TOC) was measured by using a TOC Shimadzu 5000-A.

The concentration of ions in solution was determined on a DIONEX Ionic Chromatograph equipped with a GP50 gradient pump, ED50 electrochemical detector and an IonPac AS11-HC column (4 × 250 mm) using aqueous NaOH (30 mM) as eluent and a flowrate of 1 $mL \cdot min^{-1}$.

The toxicity of samples was determined by using the Lemna minor toxicity test [21]. Glass Petri dishes containing 13 ± 2 fronds of common duckweed (L. minor) were placed under constant visible radiation (one 18 W fluorescent tube placed approximately 25 cm above the test chambers) for 96 ± 2 h in a chamber with an ambient temperature of 23 ± 1°C. Four replicates were used for each sample, which included a control without pollutant and samples taken at different reaction times. To 50 mL of solution, 0.5 mL of concentrated nutrient solutions were added. To each dish 15 mL of sample at pH 7.5 - 8 were added. Growth inhibition percentage (I) was calculated with respect to the control without pollutant according to

$$I(\%) = (100C - T)C^{-1} \tag{1}$$

where C and T are the frond number mean increments for the control and the sample, respectively.

BOD_5 was analysed using special opaque and lightproof containers with an Aqualytic BOD-Sensor. COD was determined using the recovery technique with potassium dichromate ($K_2Cr_2O_7$) and SS using the filtration technique with 0.45 μm filters [21].

2.2.3. Experimental Conditions

Prior to the degradation tests, both fungicide mixtures were filtered to remove suspended solids.

Degradation tests were conducted in 250 mL capacity Pyrex glass batch reactors.

For the heterogeneous photocatalytic degradation studies 200 mL volumes with 1 g·L^{-1} of Degussa P-25 TiO_2 or Degussa P-25 TiO_2 with 7% activated carbon were used. Aeration was maintained with an aquarium pump (EOLO AC3000 model: 2.5W power, 2 L·min^{-1} output and pressure > 0.02 MPa) and a constant stirring rate of 450 rpm. A 60W Philips Solarium HB175 equipped with four 15 W Philips CLEO fluorescent tubes with emission spectrum from 300 to 400 nm (maximum around 365 nm) was used as UV source in the degradation studies. The tests were conducted at natural pH and pH 5, as various studies have reported that photocatalytic processes are more efficient at slightly acidic pHs [22].

For the Fenton and photo-Fenton tests various concentrations of Fe^{2+} and H_2O_2 were used, with reaction volumes of 200 mL and a working pH equal to 3, adjusted with sulphuric acid. A constant stirring of 450 rpm was maintained. For the photo-Fenton reactions the Philips Solarium as described above was used as UV source. Catalase was used to remove excess H_2O_2 and to avoid interferences in the analytical and toxicity measurements.

After adsorption equilibrium was established, samples were taken to monitor the reaction in time intervals of 30 minutes or less, according to the kinetics established for each reaction. Before being measured the samples were filtered using 0.45 µm syringe filters.

3. RESULTS

3.1. Thiabendazole

3.1.1. Photocatalysis: TiO_2 and TiO_2 + AC

Figure 2 shows the results of thiabendazole degradation at natural pH (pH = 6.3) using the TiO_2based photocatalytic process. It can be seen that after 120 minutes of reaction time 95% of the thiabendazole present in the sample has been removed and after 390 minutes the amount of thiabendazole falls below the wastewater limit (0.05 mg·L^{-1}) as laid down in Canary Island (Spain) legislation.

TOC kinetics are much lower than for degradation. So, after 120 minutes only 37% of the compound had been mineralized, rising to 76% after 240 minutes.

As can be observed, the initial sample had a toxicity value equivalent to an inhibition level of 77%. This falls to non-toxic values after 120 minutes of reaction time and, though small toxicity values can be

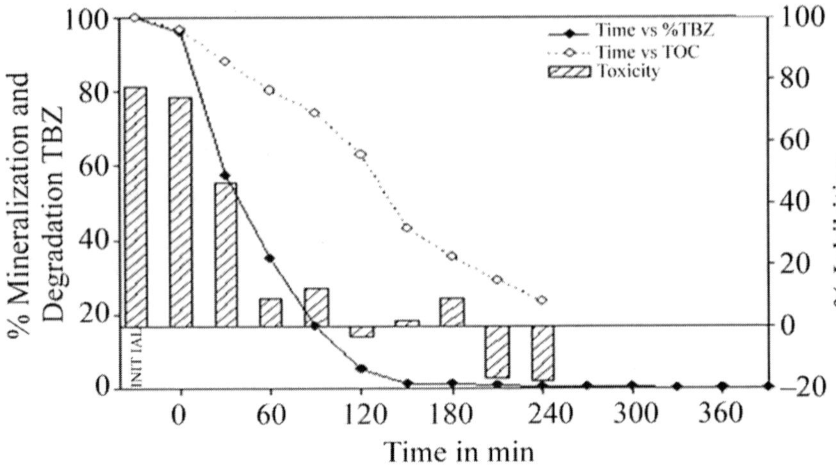

Figure 2. Thiabendazole concentration, TOC and toxicity evolution during the photocatalytic treatment of a 25 $mg{\cdot}L^{-1}$ thiabendazole aqueous solution at natural pH (6.3) using 1 $g{\cdot}L^{-1}$ of TiO_2.

detected in the following minutes, the sample becomes completely nontoxic after 210 minutes. Negative values are now given as a result of the nutritive character the sample acquires with the growth of Lemna minor.

Figure 3 shows the results for photocatalysis with 7% AC-TiO_2. **Figure 3**(a) shows the results for photocatalysis with 7% AC-TiO_2 and illumination from commencement of the test. It can be seen that the thiabendazole degrades rapidly up to 90 minutes of reaction time, when 61% of the initial concentration has been removed. However, it is not until 240 minutes that the amount of this fungicide has fallen below the limit of 0.05 $mg{\cdot}L^{-1}$. The mineralization rate falls more progressively, with a remaining TOC of 39.2% after 90 minutes and 18.1% after 240 minutes.

A fall in toxicity is observed from initial inhibition levels of 77% to 6% at 90 minutes. Toxicity ceases after 120 minutes.

Figure 3(b) underlines the adsorbent effect of activated carbon, as the reactor is left in darkness for the first hour before illumination commences. During the first 30 minutes a rapid decrease can be observed of up to 33% of the initial concentration. Thirty minutes after illumination commences 97.8% of the initial thiabendazole has been degraded, though it is not until 180 minutes of illumination that the concentration falls below the limit of 0.05 mg·L^{-1}. The same values are obtained after one hour of illumination as when illumination is used from the beginning of the reaction. This shows that the period of darkness has had no other effect than to delay the mineralization and detoxification processes. The toxicity

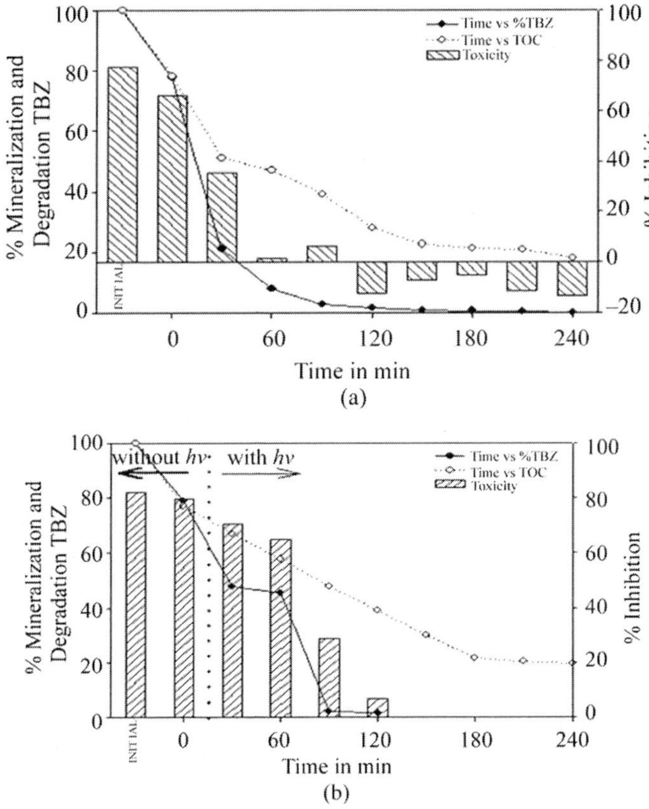

Figure 3. (a) Thiabendazole concentration, TOC and toxicity evolution during the photocatalytic treatment of a 25 mg·L^{-1} thiabendazole aqueous solution using 1 g·L^{-1} of 7% AC/TiO_2 composite as catalyst. (b) Thiabendazole concentration, TOC and toxicity evolution during the photocatalytic treatment of a 25 mg·L^{-1} thiabendazole aqueous solution using 1 g·L^{-1} of 7%AC/TiO_2 composite as catalyst but leaving the system without light for 1 hour.

level falls only slightly in the period of darkness, reaching an inhibition level of 65% at the end of that period.

3.1.2. Fenton and Photo-Fenton

Firstly, the optimal concentration of reagents (Fe^{2+}/H_2O_2) was determined for fungicide treatment by Fenton and photo-Fenton processes, namely 6.03 mg·L^{-1}/1.25 ml·L^{-1} and 5 mg·L^{-1}/0.25 ml·L^{-1}, respectively. With these concentrations, kinetic studies of thiabendazole degradation, mineralization and detoxification were then undertaken.

Figure 4(a) shows how the concentration of thiabendazole falls very quickly in the first 60 minutes of the Fenton reaction, with 98.9% of the fungicide removed and 27% mineralization. However, it is not until 240 minutes have elapsed that a value below the limit of 0.05 mg·L^{-1} is achieved, corresponding to 54.5% mineralization. Toxicity levels also respond favourably, with total detoxification occurring after 60 minutes of reaction time.

The results obtained using the photo-Fenton process (**Figure 4**(b)) show high degradation of the compound at 30 minutes (up to 99.7%). As of this point, the sample is no longer toxic. However, it is not until 90 minutes have elapsed that a concentration value below the limit of 0.05 mg·L^{-1} is achieved. At this point there is a corresponding 47% TOC removal. Though both the Fenton and photo-Fenton processes manage to lower thiabendazole content to values of 0.05 mg·L^{-1}, the photoFenton process additionally achieves a better mineralization rate of up to 80% in 250 minutes. Even so, both processes require less time for the fungicide treatment than the TiO_2 process.

3.2. Imazalil Sulphate

3.2.1. Photocatalysis: TiO_2 and TiO_2 + AC

Figure 5(a) shows the results of photocatalytic degradation with TiO_2 at the original pH of the sample,containing 15 mg·L^{-1} of imazalil sulphate. Very rapid degradation can be observed during the first 30 minutes, reaching a degradation level of 98.2%. A fungicide concentration level of 0.05 mg·L^{-1} was achieved after 60 minutes of reaction time.

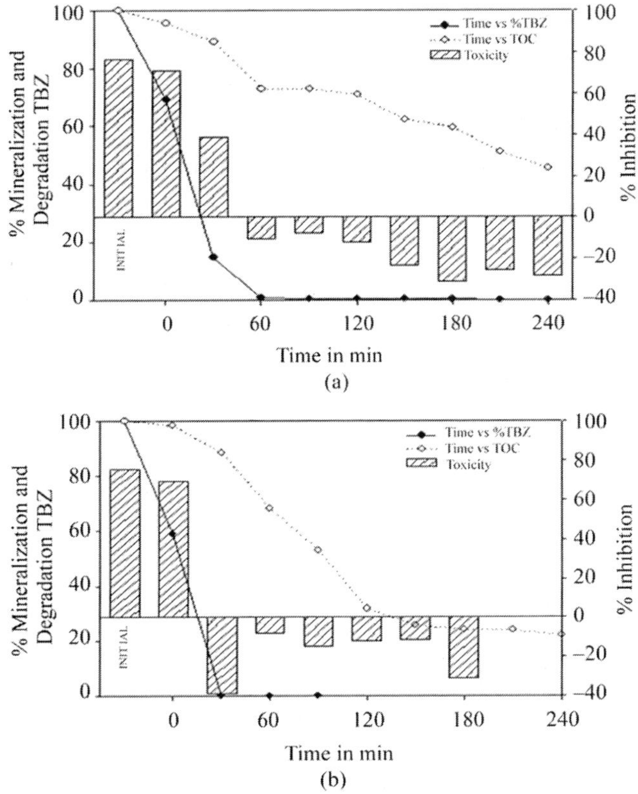

Figure 4. (a) Thiabendazole concentration, TOC and toxicity evolution during the Fenton treatment of a 25 mg·L^{-1} thiabendazole aqueous solution using 6.03 mg·L^{-1} of Fe^{2+} and 1.25 ml·L^{-1} of H_2O_2. (b) Thiabendazole concentration, TOC and toxicity evolution during the photo-Fenton treatment of a 25 mg·L^{-1} thiabendazole aqueous solution using 5 mg·L^{-1} of Fe^{2+} and 0.25 ml·L^{-1} of H_2O_2.

Toxicity levels show a significant decrease as the concentration of imazalil sulphate falls, ceasing to be toxic after 30 minutes despite a still high TOC content of 74%. Slight mineralization is observed until 60 minutes of reaction time, at which point there begins a sharp fall. This might be due to the presence of difficult-to-remove intermediates in the first stages of degradation.

Figure 5(b) shows the results when TiO_2-7% AC is used as catalyst. A very rapid fall of the fungicide concentration is observed in the first 20 minutes (94.7%), a fall which continues progressively until its total elimination after 60 minutes of reaction time.

Mineralization decreases at a slower rate. After 30 minutes of reaction time 56.6% of the initial TOC has been removed. However, as in the case of TiO_2 the sample ceases to be toxic after 30 minutes reaction time.

3.2.2. Fenton

Optimization of the concentrations of the reagents H_2O_2 and Fe^{2+} for the Fenton process in the degradation of imazalil sulphate gave values of 0.05 $mL \cdot L^{-1}$ and 6.03 $mg \cdot L^{-1}$ respectively.

Figure 6 shows the results of degradation, mineralization and detoxification when applying the Fenton process to a 15 $mg \cdot L^{-1}$ concentration of imazalil sulphate.

It can be seen how the imazalil sulphate concentration falls rapidly, with only 2.6% of the initial concentration remaining after just one minute of the process, though the fall in mineralization has only been 17.4%. It is not until 30 minutes have elapsed that an imazalil sulphate concentration below 0.05 ppm is reached. The TOC decreases slightly until the 20 minute point, with 80.7% of the initial amount remaining in solution. As of this point the decrease is much slower, with only 23.5% removed after 90 minutes of the process. Despite this persistence of degradation of the intermediates, toxicity levels tend to fall as the reaction progresses, from an initial inhibition value of 76% to 17%.

The photo-Fenton process was not applied in the study of this fungicide due to the good results obtained from the Fenton reaction on its own.

4. DISCUSSION

It can be clearly seen for thiabendazole degradation that the photo-Fenton process provided the most efficient reaction, followed by the

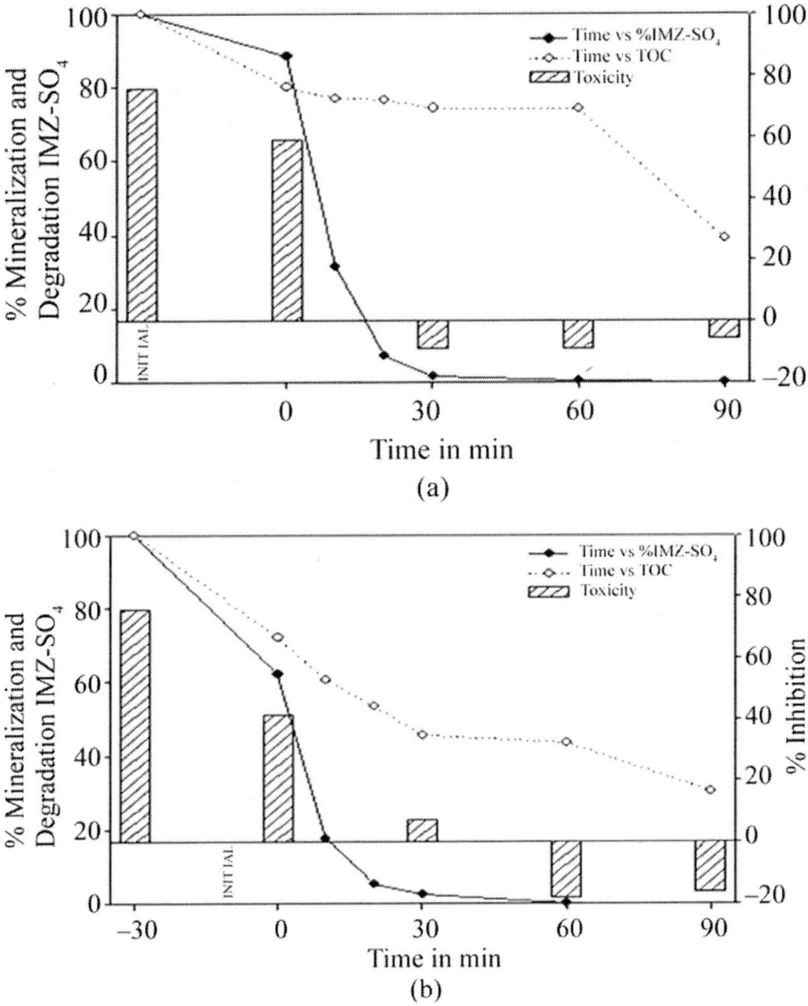

Figure 5. (a) Imazalil sulphate concentration, TOC and toxicity evolution during the photocatalytic treatment of a 15 mg·L^{-1} Imazalil sulphate aqueous solution at natural pH (5.22) using 1 g·L^{-1} of TiO_2. (b) Imazalil sulphate concentration, TOC and toxicity evolution during the photocatalytic treatment of a 15 mg·L^{-1} Imazalil sulphate aqueous solution at natural pH (5.22) using 1 g·L^{-1} of 7%AC/TiO_2 composite as catalyst.

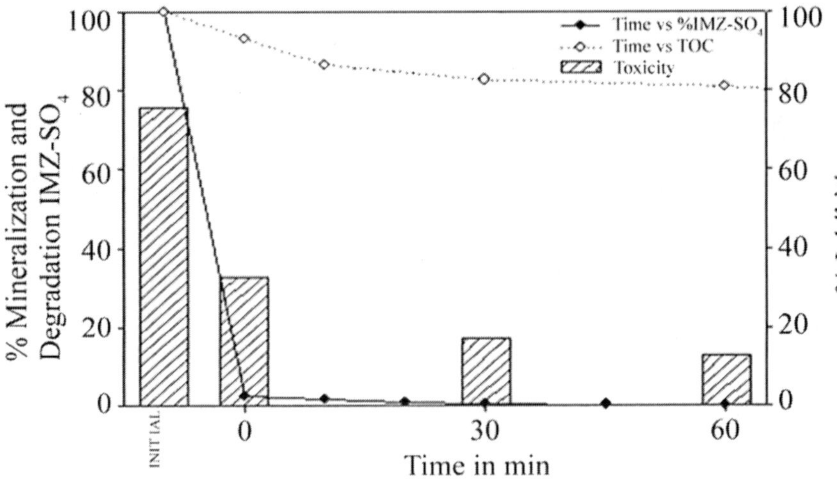

Figure 6. Imazalil sulphate concentration, TOC and toxicity evolution during the Fenton treatment of a 15 mg·L^{-1} Imazalil sulphate aqueous solution using 5 mg·L^{-1} of Fe^{2+} and 0.25ml·L^{-1} of H_2O_2.

Fenton reaction, the combination of titanium dioxide and activated carbon, and finally the least efficient reaction using only titanium dioxide as catalyst. Though all processes led to a concentration of the compound below 0.05 mg·L^{-1}, the photoFenton method was the fastest to achieve this figure. In general, inhibition of the thiabendazole degradation was observed at low concentrations. So, in the TiO_2 based photocatalytic process only 5% of the compound is degraded between 120 and 390 minutes of illumination. This figure improves in the presence of activated carbon, which has a positive effect on the photocatalytic process as it enables pollutants to get closer to the photoactive surface of the TiO_2.

The mineralization kinetic displays slower evolution than the fungicide degradation due to the intermediates generated in the course of the reaction. Evolution of the TOC shows greater efficiency in the titanium dioxide and activated carbon based process, with the least efficient process being the Fenton process.

The analysis of toxicity is of great importance as, during the degradation processes, intermediates are produced which could turn out to be more toxic than the original contaminant itself [23]. Favourable evolution of toxicity was noted for all the tested processes and for both fungicides. However, the first hour of the reaction was the most efficient, with a much faster degradation kinetic, especially in the photo-Fenton process. For this parameter, the least effective process was photocatalysis with TiO_2, which can be improved by adding activated carbon to the process as a support to the catalyst. After 120 minutes reaction time the samples cease to be toxic in all the cases despite the remaining organic carbon content.

Very high efficiency levels with short reaction times were obtained for imazalil sulphate in all the tested processes. The time periods are lower than those obtained for thiabendazole due to the fact that there was a lower initial concentration of imazalil sulphate. In the heterogeneous photocatalysis process with TiO_2, just one hour was needed to reach the concentration threshold of 0.05 $mg \cdot L^{-1}$, as opposed to the six and a half hours required for the thiabendazole. The Fenton process saw this time reduced to just 30 minutes, with additionally the need for illumination obviated.

With respect to mineralization, as for thiabendazole, the TOC showed a much more favourable evolution when a TiO_2 based catalyst was applied in combination with activated carbon than with the application of any other of the tested processes. However, the beneficial effect of the activated carbon in terms of mineralization kinetics was not so marked as for the case of thiabendazole. For its part, the Fenton process displayed very slow mineralization.

With respect to toxicity, though there was clear final detoxification of the sample with all the processes, the photocatalytic processes stand out against the Fenton process. This was probably due to traces of H_2O_2 present in the sample, as catalase was not used in the tests conducted with imazalil sulphate since it was not considered necessary due to the low concentrations of H_2O_2employed.

In general, the degradation kinetics were faster than those of mineralization. However, this did not have a negative effect on detoxification of the solution, as complete detoxification of the sample was achieved by the end of all the reactions. This would seem to indicate that the toxicity was closely related to the fungicides under study and that their degradation compounds, at the concentrations produced, were less toxic than the initial compound, or even, at reaction times greater than 60 minutes, acted as nutrients.

In summary, the Fenton process showed itself to be the most favourable option in the treatment of this type of wastewater. In addition to short degradation times, implementation of this process is the most economic from an engineering point of view.

Table 2. Parameters of the water after treatment and legally permitted levels.

Parameters	Fenton Thiabendazole	Fenton Imazalil sulphate	Legislation (RD 1620/2007)
Concentration ($mg \cdot L^{-1}$)	<0.05	<0.05	0.05
pH	7.5	7.5	5.5 - 9.5
BOD5 ($mg \cdot L^{-1}$)	<20	<20	25
COD ($mg \cdot L^{-1}$)	70.29	72.29	125
SS ($mg \cdot L^{-1}$)	9	9	35
Nitrates ($mg \cdot L^{-1}$)	0.08	0.02	10
Nitrites ($mg \cdot L^{-1}$)	0	0	
Sulphates ($mg \cdot L^{-1}$)	5.97	5.95	2000
Chlorides ($mg \cdot L^{-1}$)	2.52	2.98	2000

The concentrations of reagents used in the Fenton process are relatively low. It should be mentioned that for both thiabendazole and imazalil sulphate, the concentrations of H_2O_2 used were lower than those calculated theoretically [11], namely 0.51 $mL \cdot L^{-1}$ and 0.16 $mL \cdot L^{-1}$ for thiabendazole and imazalil sulphate, respectively. It is also known that high concentrations of peroxide can give rise to competitive reactions which produce an inhibitory effect on degradation of the substrate [14].

For this reason, the use of small quantities of this reagent is beneficial not only from an economic point of view, but also in terms of the final quality of the treated water. It has been shown that excess hydrogen peroxide generates interferences in measurements of toxicity in samples [24]. Though this particular parameter is not subject to legislation, in terms of reusing the treated water for irrigation purposes its use would not be appropriate.

Table 2 shows the results of the analysis of other chemical parameters of the wastewater after application of the Fenton process together with the legally permitted margins for water reuse for agricultural purposes. These parameters are water quality indicators [25-29] which must be fully studied and measured before treated wastewater is disposed of into the environment or reused. It can be observed in the Table that all the parameters are within the permitted margins and, therefore, this water can be reused for agricultural purposes.

5. CONCLUSIONS

The efficiency of advanced oxidation techniques has been demonstrated in the decontamination and detoxification of thiabendazole-based and imazalil sulphatebased water used in the postharvest treatment of bananas. However, Fenton and photo-Fenton processes displayed the greatest efficiency in the degradation of these fungicides in water.

The Fenton process requires lower amounts of reagents and is the simplest in terms of system design, making it the most economic from both an economic and maintenance point of view. Should further studies on a prototype or industrial scale system be instigated, it is particularly recommendable for wastewaters resulting from postharvest treatment of bananas.

Finally, particular mention should be made of the fact that, according to the legislation in force, all the processes studied in this paper resulted in a water quality of a standard suitable for reuse in irrigation.

6. ACKNOWLEDGEMENTS

The authors would like to express their thanks to COPLACA for supplying the water studied in this work, which forms part of the chemical engineering end of degree project submitted in September, 2009, at Las Palmas de Gran Canaria, entitled "Decontamination process for fungicides in water used for banana washing with advanced photochemical oxidation techniques" (translation of original title in Spanish).

We wish to thank the Spanish Ministry of Science and Innovation for their financial support through the Project CTQ2008-05961-C02-02.

7. REFERENCES

1. Coplaca, "Pudrición de Corona en el Plátano Canario," Coplaca, Canarias, 2004.
2. Boletín Oficial de Canarias (BOC), "Decreto 174/1994, de 29 Julio, por el que se Aprueba el Reglamento de Control de Vertidos para la Protección del Dominio Público Hidráulico," Boletín Oficial de Canarias, Vol. 104, No. 230, 1994, pp. 6123-6141.
3. Boletín Oficial de Canarias (BOC), "Decreto 82/1999, de 6 de Mayo, por el que se Aprueba el Plan Hidrológico Insular de Gran Canaria," Boletín Oficial de Canarias, Vol. 73, No. 73, 1999, pp. 8382-8436.
4. J. A. A. Rojas, L. F. G. Giraldo and A. F. C. Castellanos, "Fotodegradación del Pesticida Mertect Empleando FotoFenton con láMpara de luz Ultravioleta," Revista Lasallista de Investigación, Vol. 3, No. 2, 2006, pp. 19-24.
5. J. A. A. Rojas, L. F. G. Giraldo and A. F. C. Castellanos, "Fotodegradación de las Aguas Residuales con Pesticida Mertect en la Industria Bananera Empleando Fotocatálisis con Dióxido de Titanio y Lámpara de luz Ultravioleta," Revista Lasallista de Investigación, Vol. 4, No. 1, 2007, pp. 1-7.

6. J. Araña, C. G. i Cabo and C. F. Rodríguez, “Combining TiO_2-Photocatalysis and Wetland Reactors for the Efficient Treatment of Pesticides,” Chemosphere, Vol. 71, No. 4, 2008, pp. 788-794.

7. S. Malato, J. Blanco and J. Cáceres, “Photocatalytic Treatment of Water-Soluble Pesticides by Photo-Fenton and TiO_2 Using Solar Energy,” Catalysis Today, Vol. 76, No. 2-4, 2002, pp. 209-220.

8. J. B. Gálvez, S. M. Rodríguez and C. A. E. Gasca, “Purificación de Aguas por Fotocatálisis Heterogénea: Estado del Arte,” In: M. A. Blesa, Ed., Eliminación de Contaminantes por Fotocatálisis Heterogénea, La Plata, Argentina por Gráfica, 2001, pp. 51-76.

9. J. Araña, J. M. Doña-Rodríguez and E. T. Rendón, “TiO_2 Activation by Using Activated Carbon as a Support. Part II. Photoreactivity and FTIR Study,” Applied Catalysis B: Environmental, Vol. 44, No. 2, 2003, pp. 153-160.

10. J. Araña, J. M. Doña-Rodríguez and E. T. Rondón, “TiO2 Activation by Using Activated Carbon as a Support. Part I. Surface Characterisation and Decantability Study,” Applied Catalysis B: Environmental, Vol. 44, No. 2, 2003, pp. 161-172.

11. J. C. Velázquez, “Evalución Analítica y Optimización de Procesos de Oxidación Avanzada en Planta Piloto Solar,” Ph.D Thesis, Almeria University, Almeria, 2002.

12. S. H. Lin and C. C. Lo, “Fenton Process for Treatment of Desizing Wastewater,” Water Research, Vol. 8, No. 31, 1997, pp. 2050-2056.

13. Safarzadeh-Amiri, J. R. Bolton and S. R. Cater, “Ferrioxalate-Mediated Solar Degradation of Organic Contaminants in Water,” Solar Energy, Vol. 56, No. 5, 1996, pp. 439-443.

14. X. Domènech, W. F. Jardim and M. I. Litter, “Procesos Avanzados de Oxidación Para la Eliminación de Contaminantes,” In: M. A. Blesa, Ed., Eliminación de Contaminantes por Fotocatálisis Heterogénea, La Plata, Argentina por Gráfica, 2001, p. 24.

15. H. D. Burrows, L. M. Canle and J. A. Santaballa, “Reaction Pathways and Mechanisms of Photodegradation of Pesticides,”

Photochemistry and Photobiology, Vol. 67, No. 2, 2002, pp. 71-108.

16. S. Chiron, A. Fernandez-Alba, A. Rodriguez and E. Garcia-Calvo, "Pesticide Chemical Oxidation: State-of-theArt," Water Research, Vol. 34, No. 2, 1999, pp. 366-377.

17. J. L. Acero, F. J. Benitez, M. Gonzalez and R. Benitez, "Kinetics of Fenuron Decomposition by Single-Chemical Oxidants and Combined Systems," Industrial & Engineering Chemistry Research, Vol. 41, No. 17, 2002, pp. 4225-4232.

18. C. S. C. Teixeira, L. Mendes and G. Stollar, "PhotoFenton Remediation of Wastewaters Containing Agrochemicals," Brazilian Archives of Biology and Technology, Vol. 48, No. Spe., 2005, pp. 207-218.

19. S. A. S. Alvear, "Evaluación de la Aplicación en Postcosecha del Fungicida 'Pyrimethanil' via Termonebulización en el Control de Botrytis Cinerea en Manzana cv. Fuji," Agricultural Science Faculty, Agronomy School, Chile, 2003.

20. H. Kaiser, "Die Berechnung der Nachweisempfindlichkeit," Spectrochimica Acta, Vol. 3, No. 1, 1947, pp. 40- 67.

21. APHA, "Standard Methods for the Examination of Water and Wastewater," Washington DC, 2005.

22. J. Blanco Gálvez, S. Malato Rodríguez, C. A. Estrada Gasca, E. R. Bandala, S. Gelover and T. Leal. "Purificación de Aguas por Fotocatálisis Heterogénea," In: M. A. Blesa, Ed., Eliminación de Contaminantes por Fotocatálisis Heterogénea, La Plata, Argentina por Gráfica, 2001, pp. 51-76.

23. J. A. H. Melián, J. Araña and J. A. Ortega, "Comparative Study of Phenolics Degradation between Biological and Photocatalytic Systems," Journal of Solar Energy Engineering, Vol. 130, No. 4, 2008, pp. 1-7.

24. Bozzi, M. Dhananjeyan and I. Guasaquillo, "Evolution of Toxicity during Melamine Photocatalysis with TiO_2 Suspensions," Journal

of Photochemistry and Photobiology A: Chemistry, Vol. 162, No. 1, 2004, pp. 179-185.

25. Real Decreto 140/2003, "De 7 de Febrero, por el que se Establecen los Criterios Sanitarios de la Calidad del Agua de Consumo Humano," Boletín Oficial del Estado, No. 45, 2003, pp. 7228-7245.

26. Real Decreto 261/1996, "De 16 de Febrero, Sobre Protección de las Aguas Contra la Contaminación Producida por los Nitratos Procedentes de Fuentes Agrarias," Boletín Oficial del Estado, No. 61, 1996, pp. 9734-9737.

27. Real Decreto 1620/2007, "De 7 de Diciembre, por el que se Establece el réGimen Jurídico de la Reutilización de las Aguas Depuradas," Boletín Oficial del Estado, No. 294, 2007, pp. 50639-50661.

28. C. G. i Cabo, "Estrategias de Optimización de Procesos de Descontaminación de Efluentes Acuosos y Gaseosos Mediante Fotocatálisis Heterogénea," Ph.D. Thesis, Universidad de Las Palmas de Gran Canaria, Las Palmas de Gran Canaria, Spain, 2007.

29. H.-C. Su and A.-Y. Lin, "High Performance Liquid Chromatographic Determination of Imazalil Residue in Agricultural Products," Journal of Food and Drug Analysis, Vol. 11, No. 4, 2003, pp. 296-301.

8

GREEN CHEMISTRY IN PROTECTED HORTICULTURE: THE USE OF PEROXYACETIC ACID AS A SUSTAINABLE STRATEGY

Gilda Carrasco [1,*] and Miguel Urrestarazu[2]

[1]Departamento de Horticultura. Facultad de Ciencias Agrarias, Universidad de Talca, Chile

[2]Departamento de Producción Vegetal, Escuela Politécnica Superior, Universidad de Almería, Almería, 04120, Spain

ABSTRACT

Global reduction of chemical deposition into the environment is necessary. In protected horticulture, different strategies with biodegradable products are used to control pathogens. This review presents the available tools, especially for the management of protected horticultural species, including vegetables and ornamental plants. An analysis of the potential for degradable products that control pathogens and also encourage other productive factors, such as oxygen in the root system, is presented. Biosecurity in fertigation management of protected horticulture is conducted by using peroxyacetic acid mixtures that serve three basic principles: first, the manufacture of these products does not involve polluting processes; second, they have the same function as other chemicals, and third, after use and management there is no toxic residue left in the environment. The sustainability of protected horticulture depends on the development and introduction of technologies for implementation in the field.

Keywords: green chemistry; phytotoxicity; horticulture; peracetic acid; postharvest life; food industry

1. SUSTAINABLE PROTECTED HORTICULTURE AND ITS CHALLENGES

Production processes, especially those that are intensive [1,2], such as protected horticulture, degrade the environment to various extents. This one is the industry that in good part gives vegetables all over the world, due to the partial or total control of the conditions climatic of culture. The development of chemical products has permitted the food and protected horticulture industries to improve microbial safety [3]. However, the protected horticulture sector is expected to use low toxicity products that will maintain the sanitary control of vegetables grown in protected environments (greenhouses, tunnels and padding). Growers are required to combine good agricultural practices and standards of quality assurance, so food is healthy and attractive when consumed by the population [4]. This presents a global challenge for the protected horticulture industry.

One of the strategies recommended for the protected horticulture industry is the incorporation of new, environmentally friendly technologies, such as innovative green chemistry [5–7], as part of a policy-driven approach to prevent and reduce contamination. This strategy can be achieved by complying with the principles of green chemistry formulated by Anastas and Warner in 1998 (cited by [8]), which promote sustainability, efficiency and economy.

2. PRINCIPLES OF GREEN CHEMISTRY

The principles of green chemistry are key tools for achieving sustainability, as they combine the possibility of synthesis of new products with conditions for assuring that they are degradable and non-toxic to the environment [5]. These principles are:

1. To prevent the generation of waste; prevention is better than treating waste after its formation.
2. To design synthesis methods to generate new products from start to finish.
3. To design chemical syntheses that are less toxic to human health and the environment.
4. To (re)design chemical products to preserve their efficacy but reduce toxicity.
5. To use safe solvents, separating agents and catalysts and not using these when possible.
6. To recognise the energy requirements for their environmental and economical impacts. The synthesis methods should take place at room temperature and pressure.
7. To use renewable materials.
8. To avoid chemical derivates. To use safe solvents and reaction conditions.
9. To use catalytic reagents (as selective as possible) in greater proportions than the stoichiometric reagents.
10. To design biodegradable products.
11. To analyse the chemical processes in real time to avoid contamination.
12. To minimise the risk of accident spills or other releases of chemicals.

3. Strategies of Green Chemistry in Protected Sustainable Horticulture: The Use of Peroxyacetic Mixtures

The use of sodium hypochlorite is questioned and not suggested as a green chemistry alternative for its application in agriculture for pathogen control or in fertilisers. Therefore, we consider the continual use of peroxyacetic acid mixtures in liquid solutions that meet all of the principles of green chemistry outlined above.

Peroxyacetic acid ($C_2H_4O_3$), the reaction product of acetic acid (CH_3COOH) and hydrogen peroxide (H_2O_2) in liquid solution, is a very strong oxidising agent, even more powerful than chlorine or chlorine dioxide. It is translucent and colourless and produces no foam, which makes it suitable for use in irrigation facilities. Peroxyacetic acid has a very pungent smell and a pH of 2.8, and must subsequently be stabilised [8]. Peroxyacetic acid mixtures are primarily used as sanitisers in the food industry [9], since they can control deposits, odours and biofilms on surfaces in contact with food, such as fresh fruits and vegetables [10]. The virucidal power of peroxyacetic acid has also been reported [11]. The chemical reactions related to acid peroxyacetic (or peracetic acid) are:

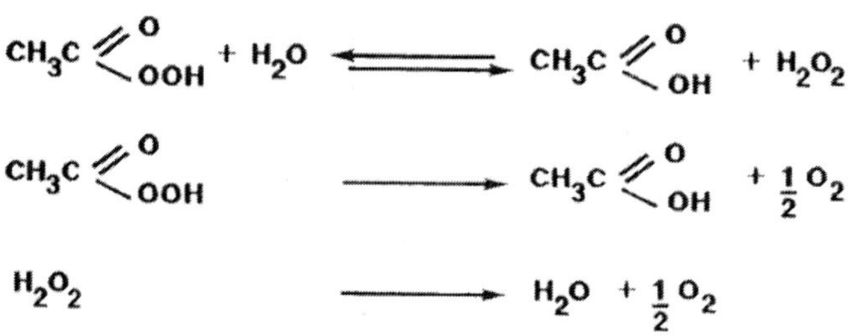

There are several registered products containing peracetic acid, hydrogen peroxide and acetic acid in different proportions as stabilised mixtures on the market. They can be used in the application of green chemistry in agriculture. None of the products formed during the degradation process (see above) are harmful. Acetic acid, hydrogen peroxide, water, and oxygen, are biodegradable, not harmful or even secondarily useful.

The specific mechanism of action of peroxyacetic acid mixtures for the control of biofilms, especially its ability to fix or remove it, is not well documented. The dissolved oxygen in a liquid medium has been described as a key factor that may be limiting for growth [12,13]. The optimum growth conditions are within a restricted range of dissolved oxygen, below which conditions are toxic and insufficient for growth.

The dose and technical management of peroxyacetic acid mixtures for the control of biofilms must be considered for efficient use and to prevent clogging of drippers or descaling. When applied in fertigation, three points should be considered and balanced for peroxyacetic acid mixture use:

1. To maintain control over pathogens when possible through its effect as a disinfectant [14];
2. The effects on secondary benefits, such as radical oxygenation on production [15]; and
3. The phytotoxic levels of each of its individual components.

Among the various functions of these peroxyacetic acid mixtures in protected horticulture and the food industry, several other uses include:

1. Disinfection of fertigation solution for control of pathogens [16];
2. Removal of drips through scaling [17,18];
3. Control of rainfall by preventing the formation of biofilms [16];
4. Increase of radical oxygenation due to the subproducts of decomposition, H_20, and 0_2, which have been evaluated in recirculation systems in soilless cultures [19];
5. Postharvest removal of natural microflora in freshly cut vegetables [10,19]; and
6. In other areas of the food industry, such as aquaculture, peroxyacetic acid mixtures control pathogens in fish by replacing other toxic products [20]. In molluscs, peroxides have also been evaluated as a promising alternative [21].

3.1. Use of Peroxyacetic Acid Mixtures for the Maintenance of Irrigation Facilities: Clogging, Environmental and Biosafety Control

Protected horticulture in south-eastern Spain constitutes a large proportion of greenhouses for the purpose of producing vegetables, and is characterized by the use of advanced irrigation and nutrition technologies. Some of the most important technological additions in this area and worldwide are localised irrigation and fertigation. The vast improvement of localised irrigation systems introduced in the past few decades has promoted more rational use of water resources, development and upgrading of marginal farming areas, as well as the use of low-quality water with high salt concentrations or even recycled water (treated water).

3.1.1. Effects on Clogging

Drip irrigation consists of localised irrigation, which provides low flows (up to 16 L h^{-1} per emitter) [22], and are ideal for securing the soil or providing an adequate amount of water. This technique enables the establishment of a wet bulb to constitute a water reserve sufficient for the development of crop growth without significant losses by deep percolation [23]. Therefore, it is possible to achieve high efficiency in water use. In the holdings of western Almería, consumption of around 6,000 $m^3\ ha^{-1}\ yr^{-1}$ can be found. By not wetting the entire soil, but only part of it, the plant concentrates its roots in the wet area so that it becomes necessary for nutrients to be taken up from the irrigated water, a practice known as "fertigation" [24]. Fertigation is commonly used in protected horticulture as it improves the benefits of drip irrigation and supports some new in relation to other disinfectants [23], maximises input, uses low-quality resources, and increases the production. This provides the agrosystem with a differential level of competitiveness and justifies the high initial cost of investment.

In places where the soil has been replaced by soilless cultivation techniques, in either the substrates only or by growing directly in a nutritious solution [24], the sustainability of the system relies on the incorporation of technologies that allow for recycling of the substrates. These substrates are present in the non-harvested plant material, which should ideally be recycled as compost [25,26]. However, to achieve this recycling management, the soil must be treated with biodegradable chemical products during fertigation and throughout the productive chain [27].

One drawback in the fertigation system is the blockage of water passage by particles, which compromises correct functioning during the time of installation [24]. The duration of the emitters must be maximised so that uniformity and efficiency of the applied water do not decrease over time; consequently, blockages, which have different origins: physical (suspended soils), chemical (formation of precipitates), biological (bacteria and algae) and their combinations, must be prevented and treated, especially in the case of buried drip irrigation systems, where visual inspection is more difficult. This concern is especially notable in the development of antimicrobial coatings, usually synthetic polymers, for internal irrigation pipes [27].

Paradoxically, clogging produces an intrinsic property, especially in the case of auto-regulation emitters. The irrigation is based in a water volume applied per unit of area and the reagent will be applied to a specific quantity of water for a given cycle. The plants that encounter a blocked emitter will not receive their volume of water, while the ones associated with a non-blocked emitter will receive more than the adequate amount of water (the emitters that function normally will be exposed to increased pressure). This phenomenon negatively influences the coefficient value of variation in the operation of the emitters [27].

The placement of several emitters per plant favours reduction of the clogging effects, because the deviation is lower. However, this greatly affects the uniformity; although two to eight emitters are installed per plant, there is an obstructed percentage of between 1 and 5% [28]. In addition, this measure is not economically viable.

The aspects mentioned above include the biosafety of the agrosystem, including water storage structures. In areas where water resources are scarce, the use of treated wastewater in urban sewage is common, and is increasingly used in protected horticulture. Although wastewater is of great interest, apart from other physical, chemical and physical-chemical problems [29], it may include a microbial level that represents a risk to human health. Under Spanish regulation, in the case of fecal coliforms, these are limited to 1,000 CFU (colony-forming units) per 100 mL for the use of irrigation water in vegetables that are to be consumed raw [30]. It is known that irrigation water is a vector of transmission of intestinal protozoa and nematodes that can cause diseases if ingested through crops irrigated with this water. In south-eastern Spain, there is contamination of irrigation water by several plant pathogens [31,32]. The presence of these pathogens, and in many cases their pathogenicity, has been demonstrated in irrigation waters [33]. Several authors, such as [34] and others, have demonstrated a high level of infectious agents, including some from the *Pythium sp,Phytophthora sp, Fusarium sp* and *Olpidium sp* families, in the irrigation waters of this region, even some reaching values above 80%. In the irrigation water, there may also be pathogenic organisms that are potentially harmful for humans and plants. These are introduced from the outside and can survive in irrigation water and adhere to biofilms to act as a reservoir, jeopardising the biosafety of the agrosystem. These biofilms are complex microorganisms and extracellular polymers fixed to a surface; they can be composed of a single or many species. Biofilms are available in any medium, provided that it is an environment with water and nutrients and that it can grow on hydrophobic or hydrophilic, and biotic or abiotic substrates [35]. This might indicate why organic matter is deposited at the interface of the water and surface, forming a layer that changes its physical properties, thereby improving the possibilities of adhesion of the bacteria. Thus, the problem of the clogging of the irrigation infrastructure by causes of biological origin is more obvious if the water used for irrigation is of residual origin. [27] classified the irrigation waters in terms of drip irrigation clogging, as indicated in Table 1.

Table 1. Effects of bacterial concentration on clogging risk [27].

Clogging risk	Number of bacteria mL^{-1}
Low	<10,000
Medium	10,000-50,000
High	>50,000

Table 2. Effect of several parameters of irrigation water on clogging risk [27].

Clogging factors	Clogging risk		
	Low	Medium	High
pH	< 7.0	7.0–8.0	>8.0
Dissolved solids (mg L^{-1})	< 50	50–100	>100
Manganese (mg L^{-1})	< 0.1	0.1–1.5	>1.5
Total iron (mg L^{-1})	< 0.2	0.2–2.0	>2.0
Hydrogen sulphide (mg L^{-1})	< 0.2	0.2–2.0	>2.0

Within the superficial uptake and use of residual wastewater, the frequent use of groundwater for irrigation is common. Most irrigation water used in arid and semiarid areas is characterised by a significant amount of dissolved bicarbonate. The natural terrains contain variable amounts of soluble salts and organic compounds, which are spread in the natural water supply. As a result, we assume that the global composition of underground water is remarkably constant, regardless of its geographic origin; the only change will be in the concentrations of each component. In regards to the amount of salt and the risk for clogging [27], established the following categories of irrigation water quality:

The dissolved ions normally precipitate by adding a precipitation germ to an initial nucleus (greater than 1 micron) to facilitate and accelerate the formation of saline precipitates that normally respond to microcrystallines. These initial geneses of the process are favoured by solid particles (under suspension) and promoted by the salinity of the fertigation water that contributes to waterlogging. The biofilm acts not only against rainfall due to its inherent biochemistry activity, but also constitutes a physical seat for the subsequent growth of precipitated salt layers [27].

3.1.2. Effects on Environmental

This interaction between the direct chemical precipitation and the synergy with the biofilms plays an essential role for two reasons. First, its direct role on clogging has already been described as an important factor. However, it also provides a "refuge" for pathogens, and thus fails to produce an effect if it is protected [27].

Traditionally, treatment of irrigation water with an acidic liquid solution was thought to be the best option for maintaining the working conditions of the irrigation system [27]. However, other options should be considered in horticulture for sustainable management.

3.1.3. Effects on Biosafety

Although the available evidence indicates that the use of chlorinated water (0.5–1.0 mg L^{-1}) is safe for crops, the use of a sodium hypochlorite solution (which releases Cl_2) has been extended in irrigation facilities to avoid clogging problems of emitters by biofilm formation. The biocidal activity of chlorine is dependent on the amount of hypochlorous acid present in the water and in contact with microbial cells. Although the mechanism of action is not entirely clear, it is predicted to inhibit enzymatic reactions and protein denaturation. These have a broad spectrum of activity (bactericidal, virucidal and sporicidal) and free chlorine retention has the most disinfectant activity [36]. The addition of sodium hypochlorite in water has the following chemical reaction:

$$NaOCl + H_2O \leftrightarrow NaOH + HOCl$$
$$HOCl \leftrightarrow H^+ + OCl^-$$

The balance of this reaction sequence depends on several factors, such as the pH of water, which has an optimum range of 6.5 to 7.5 in order to be highly efficient and stable. In addition, the temperature and presence of organic matter affect its behaviour. The hypochlorous acid reacts with the organic matter present in the washing medium and, as a result of disinfection, sub-products such as trihalomethanes, chlorine fumes [34,37], chloroform ($CHCl_3$) and bromodichloromethane ($CHBrCl_2$) are formed. These products may be carcinogenic, mutagenic, teratogenic and/or toxic gases, which have been found to be directly related to the incidence of bladder cancer and birth defects [38–41].

3.2. Strategies for the Use of Peroxyacetic Acid Mixtures to Increase the Radical Oxygenation of Vegetables Growing in Soilless Cultures

The progressive deterioration of soil covered by structures in areas of horticultural production results from exhaustion, fungal contamination and salinisation. For these reasons, the incorporation of soilless culture techniques has been proposed as an alternative [42].

In this context, the use of hydroponic methods for obtaining safe products in less time and in a reduced surface has allowed for increased cultivation acreage. These include the use of special substrates for the production of fruits and vegetables and the recirculation of systems or of direct static nutrient solutions, used especially for cultivation of leafy vegetables [24].

Despite good results being obtained, there remains a need for greater efficiency in production systems. Research needs to be conducted on the aspects that may be limiting production, for instance, those caused by poor aeration in the rhizosphere that facilitate radical formation [14]. The improvement of radical oxygenation in these systems, called "floating systems" [43] or floating table systems, through peroxyacetic mixtures was investigated. There are several methods to increase oxygen availability in the rhizosphere, from the application of air pressure to the use of peroxyacetic mixtures in irrigation water (this study). These techniques also prevent aeration problems in sensitive crops, increase the efficiency of nutrient absorption and significantly increase productivity. However, we will only consider a limited scope of their use efficiency, as they can cause serious phytotoxicity problems that may even lead to the death of the plant [44]. In the case of arugula leaf vegetables (*Eruca sativa* Mill) consumed in salads, we evaluated different doses of a peroxyacetic acid mixture in the nutrient solution, where 40 mg L^{-1} of this mixture led to a better performance of the arugula when compared with plants cultivated in substrate (perlite) without the addition of this mixture [19].

3.3. Strategies for Using Peroxyacetic Acid Mixtures to Increase the Postharvest Life of Vegetables

The industry of producing horticultural products with minimum processing is based on the selection of the best raw materials, which are cultivated under strict production systems to conserve their oganoleptic and sanitary quality with minimised losses. Sustainable strategies are needed to maintain the final product quality through reduction of the microbial load in the production chain. However, the final washing step of the product will allow for disinfection. Therefore, the use of low doses of sodium hypochlorite is still necessary, although there may be reactions with the organic matter. For this reason, the food industry is searching for sustainable alternatives [45], such as peroxyacetic acid mixtures [10]. These researchers compared a peracetic acid mix and sodium hypochlorite as disinfectants on postharvest vegetables, the results indicated that the peracetic acid mixture was better for washing fruit and improving postharvest life, because it is better for the environment (due to low toxicity) and health safety, and does not affect the taste characteristics of the fruit.

3.4. Other Uses of Peroxyacetic Mixtures in the Food Industry

As mentioned previously, peracetic acid is a strong disinfectant with a broad spectrum of antimicrobial activity [45]. Both peracetic acid and hydrogen peroxide easily degrade into oxygen and water with the potential to replace formalin in the control of fish pathogens, such as the *Multifiliis ichthyophthirius*ectoparasite. The use of peroxyacetic acid mixtures is a possible alternative to be implemented in aquaculture, according to studies conducted by [20]. Other studies conducted on molluscs aimed to replace the chlorination and chemical agents used as descaling agents, which are toxic to marine species. For mussels, the first stages of life are the most vulnerable and therefore may require reduced doses of biocides. Therefore, commercial peracetic acid has proven to be an effective alternative for species such as the*Mytilopsis leucophaeata* and embryos of *Dreissena polymorpha*. However, future studies are needed to determine the optimal concentrations of these products [21].

4. CONCLUSIONS

It is possible to incorporate green chemistry strategies in the cultural practices of protected horticulture and the food industry and to release biodegradable elements that do not pollute the environment. It is necessary to investigate the adequate doses of different peroxyacetic acid mixtures that would be incorporated into soil nutrients, substrates, or even water systems for disinfection and improvement of radical oxygenation, which would positively affect crop yield. Moreover, during postharvest, it is important to improve the quality of healthy horticultural produce demanded by the population. There are many remaining questions to be answered. What is the life cycle of these biodegradable products? How will these processes reduce the carbon footprint of horticultural products obtained from different processes that use peroxyacetic mixtures? These questions will need to be answered through future investigations considering every crop and condition of the environment.

ACKNOWLEDGMENTS

This work has been supported by the Spanish Ministry of Education and Science (Project AGL2007-64832/AGR) and the project A/7449/07 of the Spanish Ministry de Foreign Affairs and Cooperation.

REFERENCES

1. Armenta, S; Garrigues, S; de la Guardia, M. Green analytical chemistry. Trends Anal. Chem 2008, 27, 497–511.

2. García-Serna, J; Pérez-Barrigón, L; Cocero, MJ. New trends for design towards sustainability in chemical engineering: Green engineering. Chem. Eng. J 2007, 133, 7–30.

3. Warriner, K; Huber, A; Namvar, A; Fan, W; Dunfield, K. Recent advances in the microbial safety of fresh fruits and vegetables. Adv Food Nutr Res 2009, 57, 155–208.

4. Vicente, AR; Manganaris, GA; Sozzi, GO; Crisosto, CH. Chapter 5. Nutritional Quality of Fruits and Vegetables. In Postharvest Handling A Systems Approach, 2nd ed; Elsevier Inc: New York, NY, USA, 2009; pp. 57–106.

5. Kirchoff, MN. Promoting sustainability through green chemistry. Resour. Conserv. Recycl 2005, 44, 237–243.

6. Manley, JB; Anastas, P; Cue, BW, Jr. Frontiers in Green Chemistry: Meeting the grand challenges for sustainability in R&D and manufacturing. J. Clean. Product 2008, 16, 743–750.

7. Mestres, R. Green Chemistry: Views and Strategies. Environ. Sci. Pollut. Res 2005, 12, 128–132.

8. Poliakoff, M; Fitzpatrick, JM; Farren, TR; Anastas, P. Green chemistry: Science and politics of change.Science 2002, 297, 807–808.

9. Merck. Safety data sheet. Perchloric acid. Ficha de datos de seguridad. Ácido Perclórico, Available at: http://es.vwr.com/es_ES/content/servicios/consejos_practicos/ac_perclorico_merck.pdf (accessed on 24 January 2010).

10. Panreac. Peracetic acid solution 15% w/w RE. Technical Bulletin. Ácido Peracético solución 15% p/p RE. Boletín Técnico, Available at: http://www.panreac.com/new/esp/fds/ESP/X173495.htm (accessed on 18 January 2010).

11. Álvaro, JE; Moreno, S; Dianez, F; Santos, M; Carrasco, G; Urrestarazu, M. Effects of peracetic acid disinfectant on the postharvest of some fresh vegetables. J. Food Eng 2009, 95, 11–15.

12. Dell'Erba, A; Falsanisi, D; Liberti, L; Notarnicola, M; Santoro, D. Disinfection by-products formation during wastewater disinfection with peracetic acid. Desalination 2007, 215, 177–186.

13. Kornegay, BH; Andrews, JF. Kinetics of fixed film biological reactor. J. Wat. Pollut. Control. Fed 1968,40, 460–468.

14. Bungay, HR; Whalen, WJ; Sanders, WM. Microprobe technique for determining diffusivities and respiration rates in microbial slime systems. Biotechnol. Bioeng 1969, 11, 765–772.

15. Urrestarazu, M; Salas, MC; Mazuela, PC; Morales, E. Bioseguridad a través del agua de riego en la horticultura protegida. Vida Rural 2006, 239, 56–58.

16. Urrestarazu, M; Mazuela, PC. Effect of slow-release oxygen supply by fertigation on horticultural crops under soilless culture. Sci. Horticulturae 2005, 106, 484–490.

17. Urrestarazu, M; Mazuela, P; Ventura, F; Guillén, C. Beneficios de la aplicación de oxígeno en cultivos sin suelo. Agrícola Vergel: Fruticultura, horticultura, floricultura. Vida Rural 2006, 292, 195–200.

18. Eguía, E; Trueba, A; Girón, A; Río-Calonge, B; Otero, F; Bielva, C. Optimisation of biocide dose as a function of residual biocide in a heat exchanger pilot plant effluent. Biofouling 2007, 23, 231–247.

19. Verlicchi, P; Galletti, L; Masotti, L. A promising practice to reclaim treated wastewater for reuse: Chemical disinfection followed by natural systems. Desalination 2009, 247, 490–508.

20. Carrasco, G; Gajardo, JM; Álvaro, JE; Urrestarazu, M. Rocket production (Eruca sativa Mill.) in floating system using peracetic acid using as oxygen source compared with substrate culture. J Plant Nutr2010, 23.

21. Vandekinderen, I; Devlieghere, F; Van Camp, J; Denon, Q; Sánchez-Alarcon, S; Ragaert, P; De Meulenaer, B. Impact of a decontamination step with peroxyacetic acid on the shelf-life,

sensory quality and nutrient. Postharvest Biol. Technol 2009, 54, 141–152.

22. Pedersen, L; Pedersen, PB; Nielsen, JL; Nielsen, PH. Peracetic acid degradation and effects on nitrification in recirculating aquaculture systems. Aquaculture 2009, 296, 246–254.

23. Verween, A; Vincx, M; Degraer, S. Comparative toxicity of chlorine and peracetic acid in the biofouling control of Mytilopsis leucophaeata and Dreissena polymorpha embryos (Mollusca, Bivalvia). Int. Biodeterior. Biodegrad 2009, 63, 523–528.

24. Bar-Yosef, B. Advances in fertigation. Adv. Agron 1999, 65, 1–77.

25. Medina, JA. Riego Por Goteo; Teoría y Práctica; Ediciones Mundi-Prensa: Madrid, España, 2000.

26. Urrestarazu, M. Tratado de Cultivos Sin Suelo; Grupo Mundi-Prensa: Madrid, España, 2004.

27. Arvanitoyannis, IA; Varzakas, TH. Vegetable Waste Management: Treatment Methods and Potential Uses of Treated Waste; Elsevier Ltd: New York, NY, USA, 2008; Chapter 11, pp. 703–761.

28. Lazarova, V; Janex, ML; Fiksdal, L; Oberg, C; Barcina, I; Pommepuy, M. Advanced wastewater disinfection technologies: Short and long term efficiency. Water Sci. Technol 1998, 38, 109–117.

29. Anastas, PT; Warner, J. Green Chemistry Theory and Practice; Oxford Univ. Press: Oxford, UK, 2002.

30. Nakayama, FS; Bucks, DA. Water quality in drip/trickle irrigation: A review. Irrigat. Sci 1991, 12, 187–192.

31. Nakayama, FS; Bucks, DA. Emitter clogging effects on trickle irrigation uniformity. Trans. ASAE 1981,24, 77–80.

32. Segura, ML; Martín, E; Contreras, JI. Reutilización de aguas residuales urbanas para la horticultura.Horticultura 2006, 196, 16–19.

33. Gobierno de España. Boletín Oficial del Estado BOE 1999; Orden 18907, Nº 23. 1999.

34. Gill, DL. Pathogenic Pythium from irrigation ponds. Plant Disease Rep 1970, 54, 1077–1079.

35. Shokes, FM; McCarter, SM. Occurrence, dissemination and survival of plant pathogens in surface irrigation ponds in southern Georgia. Phytopathology 1979, 69, 510–516.

36. Maréchal, JP; Hellio, C. Challenges for the development of new non-toxic antifouling solutions. Int J Mol Sci 2009, 10, 4623–2637.

37. Jiménez, SMC; Domínguez, FA; Silverio, CJM. Reaction kinetics of humic acid with sodium hypochlorite. Water Res 1993, 27, 815–820.

38. O'Toole, GA. Biofilm formation as microbial development. Ann. Rev. Microbiol 2000, 54, 49–79.

39. Sapers, GM. Washing and sanitizing raw materials for minimally processed fruits and vegetables. InMicrobial Safety of Minimally Processed Foods; Novaks, JS, Sapers, GM, Juneja, VK, Eds.; CRS Press: Boca Ratón, FL, USA, 2003; pp. 221–253.

40. Reckhow, DA; Singer, PC; Malcom, RL. Chlorination of humic materials: Byproduct formation and chemical interpretations. Environ Sci Technol 1990, 24, 1655–1664.

41. Chu, I; Villeneuve, DC; Secours, VE; Becking, GC; Valli, VE. Toxicity of trihalomethanes: The acute and sucacute toxicity of chloroform, bromodichloromethane, chlorodibromethene and bromoform in rats. J. Environ. Sci. Health 1982, 17, 205–224.

42. Carpenter, LM; Beresford, SA. Cancer mortality and type of water source: Findings from a study in the UK. Int. J. Epidemiol 1986, 15, 312–320.

43. Maxwell, NI; Burmaster, DE; Ozonoff, D. Trihalometanes and maximum contaminant levels: The significance of inhalation and dermal exposures to chloroform in hosefold water. Regulat. Toxicol. Pharmacol 1991, 14, 297–312.

44. Pilotto, LS. Disinfection of drinking water, disinfection by-products and cancer: What about Australia.Austr J Public Health 1995, 19, 89–93.

45. Alarcón, A. Los Cultivos de Hortalizas Extra Tempranas; Dpto. Producción Agraria-ETSIA. Universidad Politécnica de Cartagena: Cartagena, Spain, 2005.

46. Carrasco, G; Izquierdo, J. Manual Técnico Almaciguera Flotante para la Producción de Almácigos Hortícolas; Universidad de Talca and FAO: Talca, Chile, 2005.

47. Moreno, S; Álvaro, J; Urrestarazu, M. Bioseguridad en la producción hortícola y en el consumo de alimentos. Vida Rural 2007, 259, 40–46.

48. Artés, F; Gómez, P; Aguayo, E; Escalona, V; Artéz-Hernández, F. Sustainable sanitation techniques for keeping quality and safety of fresh-cut plant commodities. Postharv. Biol. Technol 2009, 51, 287–296.

9

POSTHARVEST TECHNOLOGIES OF FRESH HORTICULTURE PRODUCE

Alejandro Isabel Luna-Maldonado[1], Clement Vigneault[2] and Kei Nakaji[3]

[1]Department of Agricultural and Food Engineering, Faculty of Agriculture, Autonomous University of Nuevo Leon, Escobedo, N.L., Mexico

[2]Department of Bioresource Engineering, Macdonald Campus, McGill University, Ste-Anne-de-Bellevue, Québec, Canada

[3]Kyushu University, Laboratory of Agricultural Ecology, Division of Agricultural Ecology, Department of Plant Resources, Fukuoka, 1Mexico Japan

1. INTRODUCTION

The actual processing of fresh horticultural produce and the reducing labor force have begun to force the development of robots that are capable of dealing with the variations inherent in the produce being handled, stored and transported. All around the world, there is increasing interest in the use of robots to replace the drastic decreasing number of farmers due to the lower birth rate and the increasing average age of the remaining farmers (Kitamura et al., 2004 & Wang, 2009). In the USA, the supply of workers available for hand harvesting is decreasing steadily and true shortages are occurring. Without enough workers when needed for a few weeks each year, a large amount of hand harvested crops will be lost. Fruit and vegetable crops need new productive harvesting technologies (Sarig et al., 2010)

In general, with horticulture applications, the variations of fresh produce in size and color as well as in the internal structure and defects such as over-ripeness and bruising damage are considered the main barriers to the extension of robotics. Robots have to cope with produce delivered with random orientation and placement, and with environmental issues such as a wide range of humidity and temperature resulting in condensation problems on electronic circuit or dew on sensors of all kinds. It requires manipulator mechanisms, controls and end-manipulators to be designed to interact with the surrounding environment, sensing techniques, mobility, and work cell development. Typical applications where imaging and electro-sensing processes can be used to guide robots include sensing general fruit quality (ElMasry et al. 2009; Li et al., 2006 & Lino et al., 2008), measuring internal quality parameters (Chen, 2008 & Li et al., 2009), or detecting mechanical (ElMasry et al., 2008) or physiological (ElMasry et al., 2009) damage, for either grading the produce accordingly or involving automatic trimming robots to remove the defect. In other words, processed items are picked from a conveyor belt and presented to a cutting or trimming device to further process the produce in order to maintain a consistency of quality in the finished produce. For example, during the inspection of lettuce heads and their core removal (Key Technology, 2010), a fresh lettuce will be scanned to detect stones and insect damage. At the same time, a two directional measurement device can take images to provide core position and orientation information. Based on these pieces of information, the lettuce will be properly aligned with a blade mounted on a robot to perform the core removal (Applied Sorting Technologies, 2010).

The above example is considered as a typical robot application in horticulture. There is also a totally different type of robot application that is often forgotten due to its relative simplicity which is to fill produce in a standard package unit such as reusable plastic container (RPC) (Vigneault and Émond, 1998). This handling process of horticultural produce is likely comparable to handling any other industrial products such as a TV set or a microwave oven. The main differences are in: the specificity of the environment due to the fact that the horticultural produce is still alive and requires special environmental conditions; the fragility of the produce to mechanical damage and its sensibility to the variations of the environmental conditions; and the length of the time the robot is used for. The particularities of environmental conditions in which the produce are handled and stored require the robot to be able to operate properly at ambient as well as in cold temperatures (close to 0ºC) and very high relative humidity (close to 100%), as well as in gas compositions that are hazardous to humans (close to 1% O2 and up to 20% CO2) in which the produce is sometimes stored (Leblanc and Vigneault, 2006); as well in a continuous alternating conditions from ambient to any other conditions. In terms of fragility of the produce to mechanical damages, it is easy to understand that horticultural produce are much more sensitive than any industrial products made from metal, plastic or wood. In terms of period of usage, the industry robots are generally used for 12 months/year period, which makes their uses generally economically interesting. However for the horticultural application, each robotic application is of any duration from a few days to the entire year depending on the type of work the robot accomplishes, the produce the robot processes and the produce production period which is produce and localization dependent, and the production system (greenhouse vs in-field production, fresh vs stored produce, cold climate vs subtropical, climacteric vs non-climacteric produce, etc). All these horticultural produce particularities must be considered during the design of any robot designated to horticultural applications.

2. HANDLING OF FRESH HORTICULTURAL PRODUCE

Postharvest technologies include the objectives of maintaining the fresh quality of the produce in terms of appearance, texture, flavor, nutritive value, etc, protecting produce, maintaining food safety, and reducing the average losses between harvest and consumption (Saraswathy et al., 2010). The time a produce is exposed to any adverse condition is generally directly proportional to the decrease of quality of any horticultural produce (Leblanc and Vigneault, 2008). Hence, the processing time is one of the most important and controllable factors affecting the quality of the produce. Any delay will generate soon or later important effects on the final quality of the produce (Vigneault et al., 2004 & de Castro et al., 2005) at the marketing step, which may result in a significant loss of marketable value (Leblanc and Vigneault, 2008). et al., 2005) at the marketing step, which may result in a significant loss of marketable value (Leblanc and Vigneault, 2008).

Another controllable factor affecting the quality of the produce is the number of times a product is manipulated or in contact with any fixed or mobile object without being protected (Vigneault et al., 2002 & Vigneault et al., 2009). Packaging systems should be designed for rapid and efficient cooling (Vigneault et al., 2002; Vigneault et al., 2004b & de Castro et al., 2005b) to prevent physical damage to the produce and to ease handling and storing processes (Vigneault et al., 2009). Finally, storage conditions of refrigerated and controlled atmosphere should be chosen and maintained to retard the deterioration of the perishable produce due to: (1) aging, ripening, softening and color changes (Hoenht et al., 2009); (2) undesirable metabolic changes and respiratory heat production(Leblanc and Vigneault, 2008) ; (3) moisture loss and the resulting wilting (Alvo et al., 2004); (4) spoilage by invasion of bacteria, fungi or yeasts (Kader, 2002); and (5) undesirable physiological processes such as sprouting (Bachmann, 2010).

2.1 Automation and robots for handling of fresh horticultural produce

To maintain the quality of horticultural produce by applying prior knowledge, manufacturers supply a range of solutions based on control, automation and robotics. For example, to reduce processing time a produce is exposed to any adverse conditions, the handling industry proposes to reduce the time between harvest and cold storage by using high speed processing systems such as display container loaders for potato, onion, carrot and other pre-packs, which are capable of obtaining a processing rate up to 80 units of 2.5 kg per minute (Abar Automation, 2011). The automated loaders are generally available for loading 800 × 600 mm footprint containers with options available for the 600 × 400 mm standard RPCs (Abar Automation, 2011). In this process, as all the main movements including package rotation are already carried out in the other industries by a robot, very few conventional technology and mechanical components are required, resulting in a very reliable system. Crate loaders are also available for pre-pack picking and placing systems that load bags, clamshell and other small package units into the 600 × 400 mm North America and Europe standard crates (Vigneault et al., 2009), 400 × 300 mm plastic crates or cases at a processing rate of more than one unit per second (Abar automation, 2009). The robot is housed with a compact space saving structure and space costs, and are fed with bulk or packed produce feeding conveyors. Outer types of packing units such as wooden crates, cardboard boxes, or RPCs are supplied to the system through conveyors either from remote or low level crate de-stackers. The remote de-stacker located on a mezzanine floor or adjacent to the pick-and-place robot separates the packing units from the top of the stack and delivers them over the packing lines. Robot de-stackers and palletizers provide a flexible and relatively low cost solution and high throughput with two or more action spots for single or multiple production lines of pallet loading.

During the picking and placing process in a packinghouse, the multitask operations allow bags, cases, boxes, RPCs and even layer sheets to be placed by a single robotic installation (Abar automation, 2009). Picking and placing is where a robot moves items from one place to another, for example from a conveyor in a production line to another in a packaging line. The robot is often guided by a vision system (Piasek, 2010). Pick and place robot cells (Fig. 1) are available for pre-packed bags or trays containing fresh or processed vegetables, salads or mushrooms ranging in weights from a few grams to 50 kg, and are capable of loading plastic crates and roll containers (Abar automation, 2011 & Robots & Robot Controllers Portal, 2010).

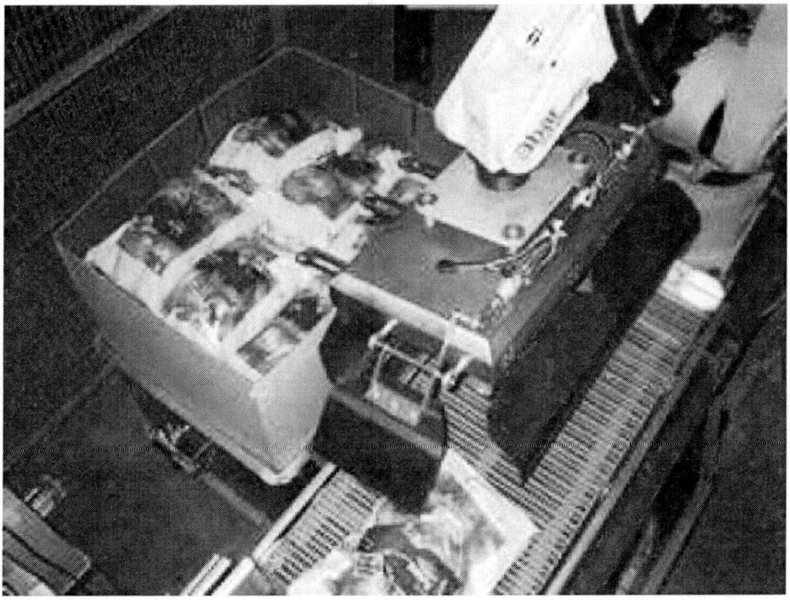

Figure 1. Pack rotation carry out by an ultra reliable (Courtesy: Abar automation, 2009).

The robotic work cells are becoming more complex. From de-moulding and assembling parts for the car industry to packaging complex objects, the robots are being asked to do faster and more complex moves. This ability to accomplish more sophisticated tasks lies in the capacity to develop more flexible end-of-arm end effecters. A customizable robot has the ability to increase functionality and repeatability as well as allow a higher percentage of automation function within the cell (KUKA Industrial Robots, 2010). For example, the crate sorting system at the Hofbräuhaus Brewery had to deal with different types of beer crates, meaning that the level of mixed packages in the system rose and manual handling became less cost-effective (Raghavan et al., 2005). In addition, the brewery wanted to spare its employees from the heavy physical work of palletizing 2,500 to 3,000 crates per day. The new solution can also be integrated into the existing equipment. The brewery is looking for a customized automation concept. Finally, the system had to be installed in a small space, while at the same time being easily accessible (Raghavan et al., 2005). This brewery example is quite similar in complexity to horticultural packinghouse where robots have to at the same time handle multiple types of fragile and light produce from conveyer to boxes, and heavy boxes to palette with care and precision, with a short time to maintain high production rate. Although the similarity in the complexity of the tasks to be accomplished is high which demonstrates the possibility of developing robots for horticulture application, the technologies are not directly applicable to horticultural produce with so many variations. Much modification to the industrial robots is required to adapt these robots for horticultural purposes. On April of this year was presented (Zheng, 2011) in China an intelligent robot (Fig. 2) ables to examine growing conditions such as temperature, humidity, light, detect disease of the vegetables and pick up the ripe ones through identifying the colour.

On the other hand, Dr. Hayashi Shigehiko in the Bio-oriented Technology Research Advancement Institution (Japan) developed the harvesting robot of strawberry in oder to keep the high quality of strawberry freshness (Fig. 3).

Various storage methods for fresh horticultural produce are used commercially (Liu, 1991 & Li, 2009). The most common and frequently used storage system is the refrigerated storage room (Raghavan et al., 2005). This storage system consists of continuous and uniform control of temperature and humidity inside of well insulated rooms or buildings (Li, 2009). The ambient storage conditions for fresh horticultural produce vary in temperature from 0 to 15° C and in relative humidity from 65 to 98% [Kader, 2002; Raghavan et al., 2005; Liu, 1991 & Li, 2009). These conditions are generally controlled by mechanical refrigeration [FAO, 2010 & Hollingum, 1999] and electric thermostat and humidistat, or using an electronic automated system based on adjusted enthalpy of the ambient air (Markarian et al., 2006).

The controlled atmosphere (CA) storage frequently used in postharvest technology of fruits and vegetables consists of controls on the concentrations of O2, CO2 and N2, within a perfectly sealed refrigerated storage room [38]. Modified atmosphere (MA) storage or packaging (MAP) is also used for horticultural produce. It consists generally of modifying gas composition of the internal atmosphere of a package unit that could be as small as 200 g (eg, for leafy vegetable) or a large

Figure 2. A robot for vegetable planting (Courtesy: Xinhuanet, Zhu Zheng, 2011)

Figure 3. A harvesting robot of strawberry (Courtesy: BRAIN, Shigehiko Hayashi, 2011).

transportable unit such as a pallet size of 250 to 1,000 kg (eg, for strawberry). The gas concentration encountered in CA and MAP are not different generally except the size of the units which is extremely different and use completely different atmosphere generation and controlling systems. The O2 gas concentration used for CA and MAP is from 21% (ambient air) to 0%, while the CO2 concentration increases from 0.003% to values sometimes reaching 15 to 20% de-pending on the produce stored. The N2 is normally used as a filling gas for both MAP and CA technologies (Raghavan et al., 2010). The ethylene (C2H4) is sometimes controlled by scrubbing such as potassium permanganate or O3 (ozone), or by flushing for better quality control of produce responding to it. The control of temperature, humidity and atmospheric composition may be performed manually, but automated systems frequently measuring the gas compositions and performing the adequate adjustment of the gas processing system also exist to achieve more accurate and uniform conditions (Raghavan et al., 2010).

3.1 Robots for storing of fresh horticultural produce

For storage of fresh produce, the robot handling system must be able to operate within the ambient conditions generally encountered in the storage rooms. At the same time, the robots have to be able to work within the conditions encountered outside of the storage room since they are continuously required to handle the produce to and from the storage room. They must then be resistant to heavy ambient humidity condensation due to this continuously changing environment. Besides the humidity condition, the basic principles used for developing industrial storage robots are basically the same for horticultural purposes. The robot must be able to circulate in a specific environment recognising fixed and moving objects consisting of rows and columns of produce, walls, doors, humans and other robots working individually or in collaboration with human and other robots.

There are two great advantages of robots used for horticultural storage purposes. The first is the robot's capacity for managing extremely large quantities of information associated with any specific produce. These pieces of information could be related to the type of produce, the origin, the harvesting date, the processing date, the initial quality, and its eventual destination, etc (Leblanc and Vigneault ,2006). The information could be easily transferred from and to the robots for managing purposes and improvement of efficiency. For example, maintaining the link between the information and the produce is essential for choosing and executing any managing types such as "the first in first out", "consumer preference", "quality based" or "shelf live" objectives. The second advantage is that in a great facility, a robot can be considered as an economical aspect in the handling the produce such as optimisation of space occupancy, organisation of traffic circulation and managing picking order (Piasek, 2010) to maximise handling efficiency and minimise produce waiting time on docks or truck loading process duration.

Another advantage of a robot, which has not been explored yet in horticulture, is the robot's capacity to work in hazardous environment such as CA room or high O3 concentration atmosphere for C2H4 control. In the case of C2H4 control, one has to understand that the optimal O3 level for the best control of C2H4 is generally too high for worker safety, which considerably limits the use of this interesting technique. For a CA room, the main problem is the fact that the quality of the produce is highly related to the time delay between harvest and storage and between storage and commercialization. It is frequent to encounter produce that has been harvested at the optimum maturity level but has to wait for days and even weeks before being under CA due to the time required to fill the CA room. The same way, it is very common to encounter CA rooms that were degassed for marketing a first load of produce and had to wait weeks and months before the last load quits the room. An airlock system has been developed for CA apple storage which can maintain the gas composition inside of the room continuously (Vigneault, 2009), reduce the waiting time before and after storage to a few hours only. However, workers have to use autonomous breading systems to enter in the airlock and transfer the produce from and to the CA room, which has to be performed with a very high precaution level. Human life can be in danger if any mistake happens (Vigneault, 2009).

4. TRANSPORTATION OF FRESH HORTICULTURAL PRODUCE

Ground transportation of fresh horticultural produce is usually by trucks and occasionally by railway (Leblanc and Hui, 2005). Overseas transportation is by ship (Tanner and Smale,2005) or plane (Thompson et al., 2004). A limited amount of high-valued produce is sometimes transported overland by air (Vigneault et al., 2009). The basic requirements for conditions during transportation are similar to those needed for storage (Vigneault, 2005), including proper control of temperature and humidity, and adequate ventilation (Hui et al., 2006 & Hui et al., 2008a). In addition to ambient condition require-ment, the produce must be immobilised by proper packaging, stacking, and bracing methods to avoid excessive movement or vibrations (Vigneault et al., 2009) without being detrimental to air circulation to maintain a uniform temperature distribution within a load (Hui et al., 2008b).

4.1 Robots for transportation of fresh horticultural produce

Loading trucks with produce can be done robots with a vision machine and integrated sensors able to detect objects or people obstructing their way and find and follow a specific path (Belforte et al., 2006). For the purpose of this handling associated with transportation, the robots should be compact enough to perform this task within a restricted work area (Garcia Ruiz et al., 2007), robust but able to perform delicate and very precise operations since the load placement within the truck greatly affects the air circulation (Vigneault , 2005; Hui et al., 2006 & Hui et al., 2008a). To perform these tasks, transportation robots coupled with forks and elevator can carry packed industrial products as well as horticultural produce (Logistic Systems Design, 2010).

5. CONCLUSION

Packaging and refrigeration prevent physical damage and decrease the physiological and pathological damages to the produce. Automation and robots are capable of speeding up the postharvest processes, decrease the period during which the perishable produce is exposed to undesirable conditions, and enhancing the facility and the safety level of human tasks for handling fresh horticultural produce. Robot palletisers could provide more than one course at the time and work within multi line production. Picking and placing robot cells adapted from industrial uses for pre-packed bags, trays or boxes containing fresh fruits or vegetables are becoming more available and able to accomplish complex tasks, but they have to be redesigned to meet with the specific requirements and environmental conditions of horticulture.

6. ACKNOWLEDGMENT

The authors would like to thank to the Mexican, Canadian and Japanese Ministries of Education for the financial support. Moreover, we thank the robot companies and institutions for the photographs.

7. REFERENCES

1. Kitamura S.; Oka K and Takeda F. (2004). Development of Picking Robot in Greenhouse Horticulture. SICE Annual Conference, pp. 3176–3179, ISBN 0780320247 Okayama, Japan

2. Wang M. What will modern agriculture bring to people? Sciences by Experts, Series 3. (2009) China Publishing Group, ISBN/ISSN 9787500084785 pp. 145– 160, Beijing, China

3. Sarig Y, Thompson JF and Brown GK. (2010). Alternatives to Immigrant Labor? The Status of Fruit and Vegetable Harvest Mechanization in the United States. 2000: 29.07.2010. Available from http://www.cis.org/articles/2000/back1200.html

4. ElMasry G, Nassar A, Wang N and Vigneault C. (2008) Spectral methods for measuring quality changes of fresh fruits and vegetables. Stewart Postharvest Review, Vol.4 , No. 4, pp. 1–3.13, ISSN 1745-9656

5. Li Z, Wang N and Vigneault C. (2006) Electronic nose and electronic tongue in food production and processing. Stewart Postharvest Review, Vol.2, No.4, pp. 7.1–7.6., ISSN 1745-9656

6. Lino ACL, Sanchez J and Dalfabro I. M. (2006). Image processing techniques for lemons and tomatoes classification. Bragantia, Vol.3, pp. 85–789, ISSN 0006-8705

7. Chen M. (2008). Non-destructive Measurement of Tomato Quality using Visible and Nearinfrared Reflectance Spectroscopy. Master degree thesis. G.S.V. Raghavan and C. Vigneault (Supervisors). Department of Bioresource

Engineering. Macdonald Campus, McGill, University. Sainte-Anne de Bellevue, Québec, Canada

8. Li Z, Wang N, Vigneault C and Raghavan GSV. (2009). Ripeness and rot evaluation of 'Tommy Atkins' mango fruit through volatiles detection. Journal of Food Engineering, Vol. 91, pp. 319–324. ISSN 0260-8774

9. ElMasry G, Wang N, Qiao J, Vigneault C and ElSayed A. (2008). Early detection of apple bruises on different background colors using hyperspectral imaging. LWT – Food science and technology, Vol. 41, No.2, pp, 337–345, ISSN 0023-6438

10. ElMasry G, Wang N and Vigneault C. (2009). Detecting Chilling Injury in "Red Delicious" Apple Using Hyperspectral Imaging and Neural Network. Postharvest Biology and Technology, Vol. 52, No.1, pp. 1–8, ISSN 0925-5214

11. Key Technology. Lettuce Core Removal System. (2010). 25.07.2010. Available from http://www.key.net/products/oncore-lettuce-core-removal-system/default.html

12. Applied Sorting Technologies. (2010). 24.07.2010. Available from http://www.appliedsorting.com.au/ index.php?~iframe1

13. Vigneault C and Émond JP. (1998). Reusable container for the preservation of fresh fruits and vegetables. Agriculture and Agro-Food Canada and Laval University. United States Patent Office, Washington, Washington DC, USA

14. Leblanc D.I. and Vigneault C. (2006). Traceability of environmental conditions for maintaining horticultural produce quality. Stewart Postharvest Review,Vol.2 , No. 2, pp. 4.1–4.10, ISSN 1745-9656

10

TISSUE BIOCHEMICAL DIVERSITY OF 20 GOOSEBERRY CULTIVARS AND THE EFFECT OF ETHYLENE SUPPLEMENTATION ON POSTHARVEST LIFE

Maria Anastasiadi[a] , Paul M. Mwangi[a] , José J. Ordaz-Ortiz[a,1] , Sally P. Redfern[b] , Mark Berry[b] , Monique S.J. Simmonds[c] , Leon A. Terry[a,]

[a]Plant Science Laboratory, Cranfield University, Bedfordshire, MK43 0AL, UK

[b]Unilever R&D Colworth, Sharnbrook, Bedfordshire, MK44 1LQ, UK

[c] Royal Botanic Gardens, Kew, Surrey, TW9 3AB, UK

ABSTRACT

The European gooseberry (*Ribes uva-crispa*) is still an understudied crop with limited data available on its biochemical profile and postharvest life. A variety of polyphenols were detected in the skin and flesh of 20 gooseberry cvs, representing mainly flavonol glycosides, anthocyanins and flavan-3-ols. In contrast, gooseberry seeds were for the first time characterised by the presence of considerable amounts of hydroxycinnamic acid glycosides tentatively identified by UPLC-QToF/MS. All cvs examined represented a good source of vitamin C while being low in sugar. Furthermore, the postharvest stability of bioactives was explored by supplementation of exogenous ethylene in air at 5 °C. Results suggest a low sensitivity of gooseberries to ethylene. The overall quality of gooseberries remained stable over two weeks, showing potential for extended bioactive life.

Keywords: *Ribes uva-crispa*; Ascorbic acid; QToF/MS; Ethylene; Storage; Bioactive life

1. INTRODUCTION

Gooseberries are deciduous shrubs being members of the *Grossulariaceae* family and genus *Ribes* like blackcurrants, redcurrants, whitecurrants and jostaberries (Bordonaba and Terry, 2011). The genus is comprised of more than 150 diverse species with currants and gooseberries being the most popular (Barney and Hummer, 2005). *Ribes uva-crispa*L. (synonym *Ribes grossularia* L.) is a European species and the most prevalent species among the gooseberries found across the world. It is native to United Kingdom, Caucasus Mountains and North West Africa (Barney and Hummer, 2005). The size of gooseberries varies as does their skin colour ranging from green to pink, red, purple, white, and yellow (Hummer and Dale, 2010). The commercial value of gooseberries is limited at present, mainly due to low demand and high cost of production especially during harvesting (Barney and Hummer, 2005 and Dale, 2000). Other influencing factors include prevalence of crop diseases such as powdery mildew (Barney and Hummer, 2005) and lack of high yielding cvs (Pluta et al., 2010).

In recent years there is a rising trend in domestic cultivation of *Ribes* berries both in Europe and other regions (Barney and Hummer, 2005 and Mitchell et al., 2011). Reasons for the increasing interest include small agricultural requirements, resistance to cold winters and the development of improved cvs with better disease resistance, colour, flavour and yield. Quality of gooseberries is primarily based on its visual, textural, organoleptic and nutritional characteristics (Terry et al., 2009). The later attribute has attracted considerable interest over the years, as some bioactive components in berries have been associated with potential health-promoting properties (Folmer et al., 2014 and Wang and Stoner, 2008).

Despite the amount of information on the qualitative and quantitative content of bioactives in berries, the nutritional quality of gooseberries has not yet been sufficiently explored due to limited commercial interest. Only a few gooseberry cvs have been studied thus far in which, polyphenols have been extracted from the fruit as a whole (Jordheim et al., 2007, Määttä-Riihinen et al., 2004, Pantelidis et al., 2007 and Wu et al., 2004). There is limited information about the spatial contribution across tissues in the final polyphenolic content of gooseberries. The main phenolic compounds reported in gooseberries include anthocyanins (Jordheim et al., 2007, Määttä-Riihinen et al., 2004,Pantelidis et al., 2007 and Wu et al., 2004), flavonol glycosides and proanthocyanins (Chiang et al., 2013, Häkkinen et al., 1999a, Mikulic-Petkovsek et al., 2012a and Russell et al., 2009).

The limited commercial value of gooseberries is also depicted in the scarce data available on the postharvest life of gooseberries and the stability of their presumed bioactive components during storage. At present, only a few reports exist on the physical, physiological and biochemical changes occurring in gooseberries during different storage conditions (Harb and Streif, 2004, Kampuse et al., 2015 and Muizniece-Brasava et al., 2015). In addition, a better understanding is needed on the role of ethylene in the postharvest life of gooseberries. Gooseberries are classified as non-climacteric fruits, although they are able to produce ethylene in low amounts (0.035–0.35 ng kg^{-1} s^{-1} at 20 °C) (Cantwell, 2002 and Thompson, 2002). The role of ethylene on *Ribes* berries, however, has not been thoroughly investigated with reports often contradictory regarding their sensitivity (Cantwell, 2002 and McKay and Van Eck, 2006).

The objective of this study was thus two-fold: to explore the biochemical profile of different tissues of a wide selection of gooseberry varieties grown in the UK, and elucidate postharvest changes in biochemistry and quality characteristics for two gooseberry varieties held for 15 days at low temperature with or without application of exogenous ethylene.

2. MATERIAL AND METHODS

2.1. Chemicals

All HPLC and LC–MS grade solvents were obtained from Fisher Scientific (Loughborough UK). (+)-catechin, (−)-epicatechin, procyanidin B1, procyanidin B2, neochlorogenic acid, caffeic acid, sinapic acid, *p*-coumaric acid, quercetin-rutinoside, quercetin-glucoside, isorhamnetin-glucoside, isorhamnetin-rutinoside were purchased from Sigma–Aldrich (Dorset, UK). Cyanidin-3-glucoside and cyanidin-3-rutinoside, were purchased from Extrasynthese (Genay Cedex, France). Metaphosphoric acid (Bioxtra ≥ 33.5%), L-ascorbic acid and D-fructose were obtained from Sigma–Aldrich (Dorset, UK). D-glucose and sucrose, were purchased from Fisher Scientific (Loughborough UK).

2.2. Plant material and sample preparation

Gooseberry fruits from 20 cvs of *R. uva-crispa* were obtained at optimum maturity, from The National Fruit Collection (Brogdale, Kent, UK) on the 13th of July 2012 for biochemical analysis (Fig. 1). Based on the biochemical results obtained, two cvs 'Careless (Kent)' and 'Scotch Red Rough' were selected and harvested again the following year (12th of July 2013) for the purposes of the postharvest trial. Approximately 100–200 berries were harvested from two plants per cv at optimum maturity stage and transported to Cranfield University in cool boxes within 2 h from collection and immediately snap-frozen in liquid nitrogen. The samples were divided into triplicates and stored at −80 °C before analysis. Each sample was further divided into two subsamples. Half of the material (approximately 100 g) was freeze-dried, the seeds were manually removed and the berries were ground into a fine powder for the extraction of phenolics and sugars. The second subsample was kept fresh frozen and powdered in a mortar grinder (RM 200, Retsch Ltd., Derbyshire, UK) with liquid N_2 and used for the extraction of ascorbic acid to avoid potential degradation of ascorbic acid during the freeze-drying process.

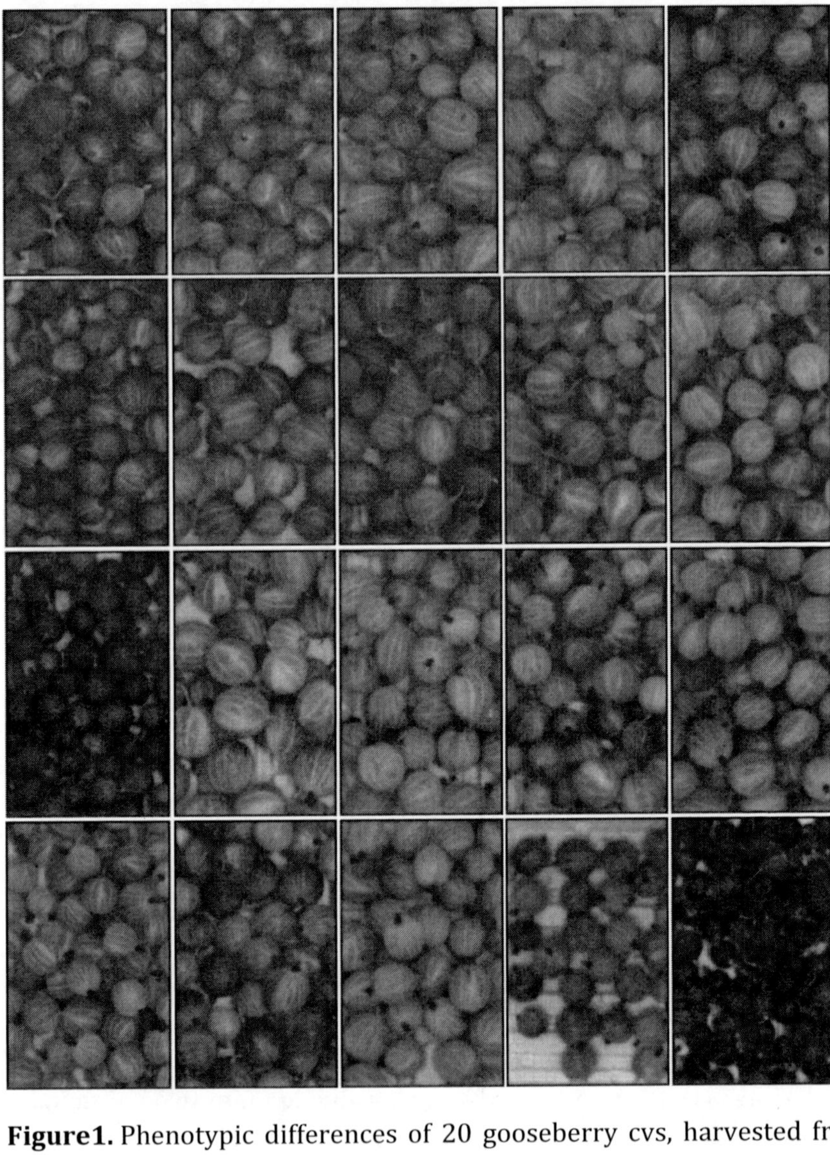

Figure1. Phenotypic differences of 20 gooseberry cvs, harvested from the National Fruit Collection in 2012. From left to right: 'Careless 3024', 'Guy seedling', 'Jolly Amylers', 'Jubilee Careless', 'Heart of Oak', 'Victoria', 'Ajax', 'Goutray's Earliest', 'Rubuste Nool', 'Lord Elco (Scotland)', 'May Dulle', 'White Eagle', 'Careless (Kent)', 'Nailer', 'Mitre', 'Cousen's Seedling', 'Lord Audley', 'Careless VT 512', 'Bedford Yellow', 'Scotch Red Rough'.

2.3. Postharvest trial

Samples from the two selected gooseberry cvs ('Scotch Red Rough' and 'Careless (Kent)') were transferred inside polystyrene boxes with ice-packs, from The National Fruit Collection to Cranfield within 3 h from harvest. Upon arrival at Cranfield, gooseberries from each cv were split into two batches. The four batches of gooseberries were placed in plastic stackable crates inside water sealed, air-tight polypropylene chambers (88 cm × 59 cm × 59 cm) fitted with two 8 × 8 cm electric fans (Nidec beta SL, RS Components Ltd., Northants, UK) to circulate the ethylene gas during treatment. Two boxes per cv were injected with 11.69 µg L^{-1} of standard ethylene gas (100% ethylene; SIP analytical) and the other two boxes were used as control samples (untreated). Temperature of the storage room was set at 5 °C and the treatment time was 24 h. The concentration of the ethylene gas in the chambers was confirmed after 30 min of injection and after 24 h of storage by withdrawing air from the chambers (including controls), using a tapped 20 mL plastic syringe. The headspace of the sampled air was injected into a gas chromatograph (GC—Model 8340, DP800 integrator, Carlos Erba Instruments, Herts, UK, analytical column, Porapak) fitted with flame ionisation detector (FID) and ethylene gas present quantified.

After the treatment was completed (24 h), every batch was split into three replicates and stored in ventilated propylene (15 cm × 22 cm × 8 cm) containers. The containers had an inlet and outlet that provided ventilation to the fruits by way of pumping air into the boxes at a flow rate of about 3.33 mL s^{-1} from a flowmeter controlled unit attached to a ICA 6000 (International Controlled Atmosphere Ltd., Kent, UK). The RH inside the boxes was maintained at 80 ± 10% by placing a beaker of water inside the containers. The RH and temperature in the boxes was continuously monitored using Tinytag Ultra 2 TGU-4500 data loggers (−95% RH, −25 to 85 °C, Gemini Data Loggers Ltd., West Sussex, UK).

2.3.1. Sampling during storage period

Upon arrival (Day 0), five gooseberries were randomly selected from each cv in triplicate. After the treatment (Day 1), five gooseberries were sampled randomly from each box (3 boxes per treatment). Sampling was thereafter repeated at regular intervals (Day 4, Day 7, Day 11, Day 13, Day 15). All samples were subjected to colour and ethylene production measurements before being snap-frozen in liquid nitrogen. The snap-frozen berries were stored at −80 °C until analysis. Prior to extraction the plant material was freeze-dried and powdered as before and analysed for individual soluble sugars and phenolics. Water content of gooseberries was calculated by recording the weight of all samples before and after freeze-drying.

2.4. HPLC-ELSD analysis of non-structural carbohydrates

Extraction of non-structural carbohydrates was performed according to a previous method with slight modifications (Terry et al., 2007). Prior to analysis, the sugars extracts were diluted (1:9, v/v) with HPLC grade water and injected into an Agilent 1200 HPLC fitted with a prevail carbohydrate ES 5 μm size of 250 nm × 4.6 mm diameter and a guard column of the same type. The mobile phase consisted of solvent A (water) and solvent B (acetonitrile) and the elution gradient was as follows: 0–15 min, 80–50% B, 15–20 min, 50–20% B, 20–25 min, 80% B. The eluted compounds were detected by evaporative light scattering detector (ELSD) and quantification was based on external calibration curves of commercial standards.

2.5. Phenolic compounds

2.5.1. Extraction of phenolic compounds

Extraction of phenolic compounds from berries and seeds was performed according to a previous method with slight modifications (Giné Bordonaba and Terry, 2008). Freeze–dried berry powder (150 mg) and freeze–dried seed powder (50 mg) were extracted with 3 mL and 1 mL, respectively, of methanol/water/HCl (70:29.5:0.5 v/v) in a shaking water bath at 20 °C for 1.5 h. The samples were vortexed every 10 min to re-suspend the solids. The extracts were subsequently filtered with 0.20 μm nylon filters and stored at −40 °C before analysis.

2.5.2. HPLC-DAD analysis of phenolic compounds and fraction collection

Analysis of flavan-3-ols, dimeric procyanidins, flavonol glycosides, anthocyanins and neochlorogenic acid in gooseberry extracts (flesh/skin) was performed on a Agilent 1200 HPLC system with a diode array detector (DAD) detector, equipped with a quaternary pump, a fraction collector, a thermostated column compartment operating at 30 °C and a cooled autosampler set at 6 °C. Separation was achieved with an Eclispse XDB-C18 column (150 mm × 4.6 mm; 5 μm, Agilent Technologies) and a 1 mm OPTI-Guard column (mobile phases consisted of 3% formic acid in water) (solvent A) and acetonitrile (solvent B) and the flow rate was 17 μL s^{-1}. The elution gradient was as follows: 0–5 min, 3–5% B, 5–10 min; 5–10% B, 10–20 min, 10–30% B; 20–25 min, 30–60% B; 25–26 min, 60–100% B, 26–30 min, 100% B, followed by 5 min re-equilibration time. The DAD detector was operating at 4 different wavelengths, 280 nm, 320 nm, 360 nm and 520 nm. Quantification was based on external calibration curves of commercial standards. Six anthocyanins detected at 520 nm for which commercial standards were not available, were quantified based on the calibration curve obtained for cyanidin-3-glucoside. Furthermore the fraction collector was used to isolate those peaks and subject them to LC/MS analysis for further elucidation.

2.5.3. UPLC/QToF/MS analysis of unknown anthocyanins

The six fractions isolated at 520 nm with UV spectra similar to the anthocyanin standards, were further analysed with an Agilent 1290 infinity UPLC system, comprised of a binary pump with a jet weaver V35 mixer, a thermostated column, set at 30 °C, and a cooled autosampler set at 6 °C. The UPLC system was coupled with an Agilent 6540 Ultra High Definition (UHD) Accurate Mass Q-TOF–MS system (Agilent Technologies) equipped with an electrospray ionization source (Agilent Dual Jet Stream). The fractions were evaporated using a miVac centrifugal vacuum concentrator (Genevac) and reconstituted with 100 μL 0.1% formic acid in water and 5 μL were injected into a ZORBAX SB-C18 Rapid resolution, HD 2.1 × 150 mm, 1.8 μm column (Agilent Technologies). Mobile phases consisted of 0.1% formic acid in water (solvent A) and 0.05% formic acid in acetonitrite (solvent B) and the flow was 6.67 μL s^{-1}. The elution gradient was as follows: 0.5–6.5 min, 5% B, 6.5–8 min; 5–15% B, 8–15.5 min, 15–20% B; 15.5–15.55 min, 20–45% B; 15.55–17.55 min, 45–100% B, 17.55–17.60 min, 100–5% B; 17.60–20.0 min, 5% B. Source conditions for the MS were set as follows: nebulizer gas (N_2) temperature 150 °C, at a flow rate of 0.13 L s^{-1} sheath gas (N_2) temperature 400 °C, at a of flow rate of 0.18 L s^{-1} and fragmentor voltage 165 V. MS spectra acquired in full scan, positive ionisation mode in the *m/z* 50–1500 range.

MS/MS experiments were subsequently carried in positive mode; to obtain the daughter ion spectrum of the selected precursor ions. Product ions were produced by collision-induced dissociation using targeted MS/MS experiments, at three different collision energies (10, 20, 40 eV) with an isolation window of 4 *m/z* (medium) for all compounds. The LC-MS system and data acquisition were controlled by an Agilent MassHunter Data Acquisition software B.04.00.

2.5.4. UPLC/QToF/MS analysis of hydroxycinnamic acid derivatives

The gooseberry extracts obtained from different tissues (skin/fresh and seeds) were further analysed with the Agilent 1290 infinity UPLC system and the Agilent 6540 UHD Accurate Mass Q-TOF-MS system described above. All extracts were further diluted (1:4) with 0.1% formic acid in water and 5 μL were injected into a ZORBAX SB-C18 column (Rapid resolution, HD 2.1 × 150 mm, 1.8 μm; Agilent Technologies) with the chromatographic method described in Section 2.5.3. Commercial standards of different phenolic compound groups were used as reference compounds (hydroxycinnamic acids, flavonols, flavan-3-ols, dimeric procyanidins).

MS/MS experiments in positive mode were subsequently carried on selected precursor ions to assist with compound identification. Product ions were produced by collision-induced dissociation using targeted MS/MS experiments, at three different collision energies (10, 20, 40 eV) with an isolation window of 4 *m/z* (medium) for all compounds.

Quantification of the hydroxycinnamic acid derivatives identified with the above method was based on external calibration curves of the corresponding aglycons (*p*-coumaric acid, caffeic acid, ferulic acid, sinapic acid).

2.6. Extraction and quantification of total ascorbic acid

The total ascorbic acid of the 20 gooseberry cvs harvested in 2012, was determined as total vitamin C using a previously described method with slight modifications (Planchon et al., 2004). The method determined the total vitamin C in the gooseberries as the sum of ascorbic acid and its oxidised form dehydroascorbic acid. Briefly, 500 mg of fresh frozen gooseberries were mixed with 2.5 mL of degassed metaphosphoric acid solution (0.01 M) and the mixture was extracted in a shaking water bath at 25 °C for 10 min. The extract was filtered through 0.20 μm nylon filters and immediately injected in the HPLC. Ascorbic acid was monitored at 248 nm and the total vitamin C was quantified using an external calibration curve of L-ascorbic acid. The assay was performed in triplicate.

2.7. Colour measurement

The colour of gooseberries was assessed using an 8 mm aperture handheld tristimulus colourimeter (Minolta CR- 400, Osaka, Japan). The mean colour space coordinates of Lightness (L^*), Chroma (C^*) and Hue angle ($H°$) were taken from three positions on each gooseberry for every sampling point.

2.8. Ethylene production measurement

The ability of the gooseberries to produce ethylene gas was measured with an ETD-300 Ethylene Detector system (Sensor Sense BV, Nijmegen, The Netherlands). In order to determine the amount of ethylene in the cuvettes after 1 h incubation, air was pumped at a continuous flow rate of 1.1 $\mu L\ s^{-1}$. The air from the cuvettes was passed through $CO_2$and water scrubbers. Measurements from every cuvette were made for a period of 12 min and expressed in $ng\ kg^{-1}\ s^{-1}$. The equipment was equilibrated before measurement by measuring an empty cuvette for 1 h. A baseline measurement of 3 min was maintained between the samples to eliminate cross contamination.

2.9. Statistical analysis

Analysis of variance (one-way and two-way ANOVA) was performed using Genstat for Windows, Version 14 (VSN International Ltd., Herts., UK). The differences between means of data were compared through Least Significance Difference (LSD) and they were considered to be statistically significant at the 95% confidence level $(p \leq 0.05)$. Logarithmic transformations were employed where needed in order to ensure the assumption of equal variability.

3. RESULTS AND DISCUSSION

3.1. Phenolic composition of gooseberries

3.1.1. Characterization and quantification of non-coloured compounds in skin/flesh

Gooseberries (flesh/skin) contained a variety of phenolic compounds including flavonol glycosides, flavan-3-ols, dimeric procyanidins, and hydroxycinnamic acid derivatives (Table 1 and Table 2). The major class of polyphenols present in all cvs was flavonol glycosides with the rutinosides of quercetin and isorhamnetin being the predominant glycosides. Quercetin and isorhamnetin glucoside on the other hand were found in much lower amounts. Mikulic-Petkovsek et al. (2012a) also observed that quercetins in gooseberries existed as glucosides and rutinosides and that these are the main form in which they exist in berries from *Glossulariaceae* family. The same authors have also reported high concentration of isorhamnetin-3-rutinoside in gooseberries (73.9 mg kg^{-1}in red and 48.3 mg kg^{-1} in white gooseberries for FW) when compared to other common berries such as bilberry, blackcurrant and strawberry. Similarly, Määttä-Riihinen et al. (2004) have also reported high concentration of isorhamnetin glycosides in gooseberries. Our results are in good agreement with these previous reports, although the concentration range of quercetin glycosides among the 20 gooseberry cvs analysed within this study were higher than those previously reported (Häkkinen et al., 1999a,Häkkinen et al., 1999b, Määttä-Riihinen et al., 2004 and Mikulic-Petkovsek et al., 2012a). These differences could reflect that, to date, only a limited number of gooseberry cvs have been studied. However, the overall trend is in line with Määttä-Riihinen et al. (2004)observation that quercetins are high in berries from *Glossulariaceae* family. Other flavonol glycosides reported in gooseberries, include glycosides of myricetin and kaempferol, typically present in the form of rutinoside representing minor constituents of the flavonol content of gooseberries (Määttä-Riihinen et al., 2004 and Mikulic-Petkovsek et al., 2012a). Furthermore Mikulic-Petkovsek et al. (2012a), identified the presence of small amounts of lacitrin and syringetin rutinoside in a white gooseberry cv and other berry species.

Table 1. Flavonoid composition ($mg\ kg^{-1}$) of the flesh/skin of 20 gooseberry cvs on a fresh weight (FW) and dry weight (DW) basis (values inside brackets).[a]

Cultivar	(+)-Catechin		(−)-Epicatechin		ProcB1[b]		ProcB[c]		Q-3-rut[d]		Q-3-gluc[e]		Isorh-3-rut[f]		Isorh-3-gluc[g]	
Scotch Red Rough	20.1	(112.8)	4.2	(23.7)	8.8	(49.5)	2.5	(14.3)	102.5	(767.9)	13.7	(102.7)	81.0	(606.4)	2.7	(20.4)
Lord Audley	26.5	(224.8)	2.2	(18.7)	9.3	(79.2)	2.4	(20.2)	78.2	(585.6)	7.0	(52.8)	38.3	(287.2)	1.6	(12.3)
Heart of Oak	44.3	(264.5)	2.4	(13.9)	12.9	(75.4)	2.3	(13.6)	55.3	(414.1)	7.1	(52.8)	26.8	(200.4)	1.2	(8.7)
Nailer	37.0	(219.3)	2.2	(13.0)	13.1	(70.3)	2.4	(14.4)	49.9	(373.8)	4.8	(35.9)	29.3	(219.1)	1.4	(10.1)
Jolly Amylers	56.5	(364.0)	2.4	(15.6)	16.4	(105.4)	2.3	(15.0)	25.8	(192.9)	3.4	(25.3)	24.0	(179.6)	1.7	(12.7)
Bedford Yellow	46.6	(326.2)	2.7	(18.5)	18.4	(128.5)	4.5	(31.2)	37.5	(280.6)	2.9	(21.9)	13.9	(104.0)	0.9	(6.6)
Cousen's Seedling	14.0	(112.6)	2.0	(16.1)	10.3	(82.5)	3.9	(31.4)	44.7	(334.5)	5.0	(37.3)	40.6	(304.2)	2.5	(18.6)
Careless (Kent)	10.9	(83.2)	1.0	(7.4)	7.0	(53.7)	1.0	(7.8)	43.1	(322.8)	5.3	(39.9)	28.7	(215.0)	2.3	(17.0)
Victoria	34.6	(285.2)	2.3	(18.8)	12.5	(102.2)	2.6	(21.0)	25.7	(192.4)	3.0	(22.5)	16.4	(123.2)	0.5	(3.9)
White Eagle	32.6	(209.6)	2.1	(13.4)	12.4	(79.3)	2.6	(16.8)	16.2	(121.4)	3.3	(25.1)	24.5	(183.2)	2.2	(16.6)
Jubilee Careless	28.5	(192.9)	1.7	(11.6)	10.5	(71.2)	1.6	(10.9)	24.6	(184.1)	2.5	(18.6)	19.5	(145.9)	1.2	(9.3)
Goutray's Earliest	29.6	(258.0)	3.0	(25.8)	10.4	(90.9)	3.7	(32.1)	20.6	(154.4)	3.0	(22.4)	18.1	(135.2)	1.4	(10.6)
May Dulle	15.1	(106.9)	2.6	(18.5)	9.7	(68.6)	1.9	(13.7)	30.6	(229.5)	5.8	(43.5)	21.4	(160.0)	1.5	(11.0)
Mitre	17.8	(126.5)	1.3	(09.5)	8.2	(58.2)	1.2	(8.9)	33.9	(254.0)	3.7	(27.6)	18.8	(141.1)	1.2	(8.9)
Lord Elco	40.1	(320.3)	1.4	(10.9)	13.0	(103.3)	1.7	(13.6)	9.6	(71.6)	0.6	(4.1)	18.3	(137.3)	1.4	(10.9)
Rubuste Nool	30.1	(206.1)	1.6	(10.9)	11.1	(76.0)	1.5	(10.0)	18.1	(135.9)	1.9	(14.5)	18.3	(137.3)	1.4	(10.3)
Careless 3024	26.9	(201.5)	2.3	(17.0)	10.7	(80.4)	2.4	(17.9)	20.2	(150.9)	1.4	(10.6)	15.0	(112.1)	0.6	(4.4)
Careless (VT)	19.9	(130.7)	1.5	(9.7)	8.3	(55.2)	1.5	(10.1)	25.4	(190.3)	2.6	(19.4)	16.8	(125.8)	1.2	(8.8)
Ajax	19.4	(132.1)	1.7	(11.5)	10.3	(69.8)	1.9	(13.2)	19.4	(145.6)	2.3	(16.9)	16.1	(120.8)	1.3	(9.6)
Guy seedling	8.8	(44.7)	1.8	(9.1)	8.7	(43.8)	3.4	(17.2)	15.4	(115.1)	2.6	(19.2)	29.0	(217.4)	2.0	(15.1)
Mean	28.0	(196.1)	2.1	(14.7)	11.1	(77.2)	2.4	(16.7)	34.8	(260.9)	4.1	(31.2)	26.0	(192.8)	1.5	(11.3)
LSD[h]	8.25	(53.48)	0.49	(2.99)	2.27	(12.33)	0.48	(2.77)	16.34	(122.38)	2.10	(15.9)	8.40	(68.4)	0.50	(3.2)

[a] The results are presented in descending order according to the sum of flavonoids for FW.
[b] ProcB1 = procyanidin B1.
[c] ProcB2 = procyanidin B2.
[d] Q-3-rut = quercetin-3-rutinoside.
[e] Q-3-gluc = quercetin-3-glucoside.
[f] Isorh-3-rut = isorhamnetin-3-rutinoside.
[g] Isorh-3-gluc = isorhamnetin-3-glucoside.
[h] LSD = least significant difference at the 95% confidence level.

Flavan-3-ols and dimeric procyanidins were present in gooseberries in small to moderate amounts with (+)-catechin being the predominant with an average concentration of 28.0 $mg\ kg^{-1}$ (FW). These results are comparable with previous reports, with the exception of (+)-catechin which was in concentrations higher than that previously reported (Häkkinen et al., 1999b). Very few reports are available describing the levels of procyanidins in gooseberries, however, according to Wu et al. (2004), procyanidins and prodelphinidins are present in gooseberries mainly in polymeric form.

Table 2. Hydroxycinnamic acid derivative composition (mg kg^{-1}) of the flesh/skin of 20 gooseberry cvs on a FW and DW basis (values inside brackets).[a]

Cultivar	Coumaroyl-hexose1		Coumaroyl-hexose2		Caffeoyl-hexose		Feruloyl-hexose1		Feruloyl-hexose2		Neochlorogenic acid	
Guy Seedling	45.5	(229.8)	15.3	(77.3)	39.9	(201.6)	2.3	(11.9)	11.8	(5.97)	10.4	(52.6)
Heart of Oak	41.5	(241.5)	18.2	(105.8)	33.4	(194.2)	1.5	(8.8)	8.2	(4.93)	7.7	(45.4)
Nailer	33.3	(198.8)	12.8	(76.5)	33.5	(199.8)	1.6	(9.9)	4.7	(2.88)	9.1	(54.5)
Jolly Amylers	38.2	(245.4)	16.7	(107.2)	24.0	(159.7)	1.7	(10.9)	7.0	(4.53)	3.8	(24.4)
Bedford Yellow	35.1	(244.5)	18.3	(127.9)	16.5	(114.7)	1.3	(8.9)	3.4	(2.44)	12.4	(87.0)
Rubuste Nool	32.3	(221.1)	13.3	(91.0)	24.9	(169.9)	2.8	(18.8)	6.9	(4.74)	4.2	(28.7)
Lord Elco	35.9	(286.0)	15.5	(123.7)	20.6	(164.6)	1.9	(15.2)	6.2	(4.99)	3.8	(30.3)
White Eagle	26.4	(168.8)	7.5	(48.0)	29.2	(186.4)	2.1	(13.7)	6.2	(3.97)	3.2	(20.8)
Careless VT	28.5	(218.9)	11.6	(88.9)	21.6	(166.0)	1.4	(11.0)	3.6	(2.76)	0.7	(5.4)
Scotch Red Rough	8.6	(48.5)	4.9	(27.3)	26.1	(145.6)	4.5	(25.0)	12.0	(6.68)	11.7	(66.8)
Mitre	25.6	(182.0)	9.3	(66.1)	20.8	(147.2)	1.4	(10.2)	3.4	(2.43)	0.8	(5.6)
Cousen's Seedling	26.5	(212.1)	9.6	(76.1)	11.2	(89.4)	2.0	(16.0)	5.5	(4.46)	3.4	(27.9)
Careless (Kent)	27.1	(179.7)	10.3	(68.1)	14.3	(94.8)	1.5	(10.3)	3.9	(2.58)	0.7	(4.8)
Careless 3024	17.2	(129.0)	7.2	(54.2)	20.0	(150.1)	1.5	(11.2)	4.5	(3.38)	5.6	(41.8)
Lord Audley	19.3	(163.7)	6.1	(51.2)	15.0	(126.6)	1.3	(11.2)	3.4	(2.84)	8.8	(73.9)
Jubilee Careless	27.6	(186.2)	10.1	(68.0)	11.4	(76.3)	1.1	(7.5)	2.6	(1.78)	0.7	(4.8)
May Dulle	6.4	(45.7)	5.5	(39.5)	22.3	(158.8)	2.3	(16.1)	5.1	(3.64)	8.4	(69.1)
Victoria	14.5	(118.6)	6.9	(56.0)	17.0	(138.4)	1.0	(8.2)	3.8	(3.04)	5.1	(42.2)
Goutray's Earliest	12.3	(107.2)	5.3	(46.7)	17.8	(155.5)	1.1	(9.6)	3.4	(2.95)	6.7	(58.1)
Ajax	17.2	(116.5)	5.8	(39.4)	13.5	(91.7)	1.4	(9.6)	4.0	(2.70)	3.9	(26.5)
Mean	26.0	(177.2)	10.5	(71.9)	21.7	(146.6)	1.8	(12.2)	5.5	(3.68)	5.6	(38.5)
LSD	5.31	(24.87)	2.59	(12.36)	5.75	(29.94)	0.51	(3.63)	0.98	(0.532)	1.61	(11.50)

[a] The results are presented in descending order according to the sum of hydroxycinnamic acid derivatives for FW.

Of the phenolic acids, only hydroxycinnamic acid derivatives were detected in low to moderate amounts in the skin and flesh of gooseberries. The presence of neochlorogenic acid (*trans*-5-*O*-caffeoylquinic acid) was verified through the use of a commercial standard and accurate mass measurements. Several hydroxycinnamic acid glycosides were also tentatively identified with QToF/MS/MS, based on accurate mass measurements and the fragmentation pattern acquired with MS/MS in three different voltages (10, 20, 40 eV) which matched the fragmentation pattern obtained for the corresponding aglycons. The most abundant fragments for each compound are presented in Table 3. The sugar moiety was a hexose and the most abundant pseudo-molecular ion in positive mode was in the form of a $[M-Na]^+$ adduct. The following glycosides were identified in gooseberry skin/flesh extracts: *p*-coumaric, caffeic, ferulic and sinapic acid hexosides (Table 3).

Table 3. Tentative identification of anthocyanins and hydroxycinnamic acid glycosides in gooseberry cvs with QToF/MS/MS.

Peak	Ret. time (min)		*m/z*	Fragments MS/MS	Tentative identification	Molecular formula
Anthocyanins						
3	7.13	[M+]	463.1303	301.0995	Peonidin-3-glucoside	$C_{22}H_{23}O_{11}$
3	7.27	[M+]	419.1190	287.055	Cyanidin-3-xyloside	$C_{20}H_{19}O_{10}$
4	7.42	[M+]	609.2129	301.0994	Peonidin-3-rutinoside	$C_{28}H_{33}O_{15}$
6	10.11	[M+]	595.1447	287.055	Cyanidin-glycoside	
7	10.36	[M+]	595.1480	(287.0814, 147.0570, 163.1326)	Cyanidin-coumaroyl-glucopyranoside	$C_{30}H_{27}O_{13}$
8	10.96	[M+]	609.1594	301.705	Peonidin-glycoside	$C_{28}H_{33}O_{15}$

Ret. time (min)		*m/z*	Fragments MS/MS	Tentative identification	Molecular formula
Hydroxycinnamic acid derivatives					
5.15, 6.10	[M − Na]+	349.0893	(165.0545, 147.0570)	Coumaric acid hexoses	$C_{15}H_{18}O_8$
4.20	[M − Na]+	365.0862	(181.0508, 163.0402, 145.0283, 135.0437, 117.0333, 89.0383)	Caffeic acid hexose	$C_{15}H_{18}O_9$
4.86, 6.42	[M − Na]+	379.0999	(195.0651, 177.0546)	Ferulic acid hexoses	$C_{16}H_{20}O_9$
5.62	[M − Na]+	409.1109	(225.076, 207.0652, 175.0391)	Sinapic acid hexose	$C_{17}H_{22}O_{10}$

Among the hydroxycinnamic acid glycosides detected in gooseberry flesh/skin extracts, coumaric and caffeic acid glycosides were the most abundant, followed by ferulic acid glycosides (Table 2). Sinapic acid glycosides were also detected in trace amounts (Table 3). 'Guy Seedling' was the cv with the highest content in hydroxycinnamic acid glycosides with a sum of 114.8 mg kg^{-1} (FW), while 'Scotch Red Rough' cv which exhibited the highest flavonoid content among all cvs, had low levels of hydroxycinnamic acid glycosides (56.1 mg kg^{-1}). The levels of hydroxycinnamic acid glycosides reported here are in good agreement with those reported in berries from other species in theGrossulariaceae family (Määttä-Riihinen et al., 2004). Neochlorogenic acid was present in small amounts in gooseberry flesh/skin with a mean concentration of 5.6 mg kg^{-1} (FW) in 'Scotch Red Rough'. Its concentration was almost 2-fold higher in coloured varieties when compared to green varieties. This is the first report of neochlorogenic acid in gooseberries. The presence of several phenolic acids has been previously reported in the literature including protocatechuic acid, *p*-hydroxybenzoic acid, *p*-coumaric acid, ellagic acid, ferulic acid, chlorogenic acid, sinapic acid, caffeic acid, syringic acid, vanillic acid and gallic acid (Filipiak-Szok et al., 2012, Mattila et al., 2006 and Russell et al., 2009). It is notable that none of the above phenolic acids were detected in free form in gooseberries in our study and no hydroxybenzoic acids were detected in the berries of the gooseberry cvs analysed. Similarly Määttä-Riihinen et al. (2004) reported the presence of hydroxycinnamic acids in gooseberries and other berry species in conjugated form, mainly as caffeoyl- and *p*-coumaroylglucosides.

3.1.2. Phenolic profile of gooseberry seeds

Unlike the flesh/skin extracts which contained a variety of different phenolic compounds, the only type of phenolic compounds detected in gooseberry extracts of seeds were hydroxycinnamic acid derivatives, which were tentatively identified as glycosides of caffeic, coumaric, ferulic and sinapic acids as described in Section 3.1.1. Two distinct peaks with the same spectral characteristic attributed to coumaric acid glycosides and ferulic acid glycosides were present in all seed extracts, indicating the presence of two isomers of coumaric acid and ferulic acid glycosides (Table 3). Their concentrations are presented in Table 4. Gooseberry seeds were observed as being particularly high in phenolic acid glycosides, with concentrations being 10 times higher compared to the flesh/skin extracts. There was some variability among different varieties with the highest concentrations observed for 'Heart of Oak'. The qualitative profile was similar to the flesh/skin extracts, with coumaric and caffeic acid glycosides being the most abundant, followed by ferulic and sinapic acid glycosides. No phenolic acids were present in free form in the seed extracts, as it was confirmed by comparison with the corresponding aglycon standards. It is also notable that neochlorogenic acid, which was detected in the flesh/skin extracts, was absent in gooseberry seeds. Other compounds detected in the gooseberry seed extracts included trace amounts of (+)-catechin and (−)-epicatechin, quercetin-3-rutinoside, isorhamnetin-3-rutinoside and cyanidin-3-rutinoside and cyaniding-3-glucoside in some red cvs.

To our knowledge no other reports exist on the phenolic composition of gooseberry seeds, but some comparisons can be made with other *Ribes* species. Gođevac et al. (2012) have tentatively identified 35 compounds in black, red and white currant seed extracts, including phenolic acids derivatives such as sinapoyl glucoside and chlorogenic acid and several phenolic acids in free form, small amounts of flavonols and flavonol glycosides as well as two nitrile containing derivatives of *p*-

Table 4. Concentration of hydroxycinnamic acid glycosides (g kg^{-1}) in the seeds of 20 gooseberry cvs on a DW basis.[a]

Cultivar	Coumaroyl-hexose1	Coumaroyl-hexose2	Caffeoyl-hexose	Feruloyl-hexose1	Feruloyl-hexose2	Sinapoyl-hexose	ΣHCAD[b]
Heart of Oak	4.55	1.09	6.65	0.15	1.10	0.20	13.74
Guy seedling	2.26	0.64	4.19	0.12	0.64	0.19	8.04
Victoria	1.85	0.67	3.99	0.19	0.68	0.33	7.71
Careless VT	2.21	0.75	3.03	0.15	0.76	0.31	7.21
Nailer	1.77	0.64	3.33	0.12	0.65	0.33	6.84
Mitre	1.81	0.65	3.26	0.15	0.66	0.28	6.81
Jolly Amylers	1.70	0.57	3.23	0.12	0.58	0.37	6.57
Bedford Yellow	1.97	0.76	1.74	0.17	0.76	0.24	5.64
May Dulle	1.35	0.46	2.47	0.15	0.47	0.18	5.08
Ajax	0.91	0.35	2.89	0.12	0.36	0.30	4.93
Jubilee Careless	1.52	0.51	1.99	0.13	0.51	0.25	4.91
Careless 3024	1.29	0.50	1.95	0.14	0.50	0.31	4.69
Careless (Kent)	1.55	0.49	1.82	0.11	0.49	0.17	4.63
Cousen's Seedling	1.53	0.45	1.75	0.16	0.45	0.29	4.63
White Eagle	1.14	0.31	2.30	0.10	0.32	0.32	4.49
Goutray's Earliest	0.82	0.33	2.48	0.13	0.33	0.20	4.29
Rubuste Nool	0.98	0.28	1.81	0.08	0.28	0.25	3.68
Lord Elco	0.99	0.30	1.60	0.10	0.30	0.31	3.60
Lord Audley	0.71	0.24	1.75	0.11	0.24	0.15	3.20
Scotch Red Rough	0.61	0.21	1.72	0.16	0.21	0.05	2.96
Mean	1.58	0.51	2.70	0.13	0.51	0.25	5.68
LSD	0.356	0.15	0.917	0.036	0.151	0.077	1.393

[a] The results are presented in descending order according to the sum of hydroxycinnamic acid derivatives for DW.
[b] ΣHCAD = Sum of hydroxycinnamic acid derivatives.

coumaric and ferulic acid. In contrast Bakowska-Barczak et al. (2009), have reported flavonols as the main phenolic group present in blackcurrant seed residue in the form of kaempferol, myricetin and quercetin glycosides. The same authors also reported the presence of amounts of *p*-coumaric acid along with small amounts of *p*-coumaroyl-glucoside in the seeds (0.487 g kg^{-1} and 0.179 g kg^{-1}, respectively, for DW), although the presence of the aglycon was possibly a degradation product (Bakowska-Barczak et al., 2009). Lu et al. (2002) isolated six glucosides of quercetin, kaempferol and myricetin from blackcurrant seeds along with two nitrile containing derivatives of *p*-coumaric acid. Furthermore, Lu and Foo (2003) isolated and characterised by ^{13}C NMR two phenolic acid conjugates namely 1-*p*-coumaroyl-β-D-glucopyranoside and 1-cinnamoyl-β-D-glucopyranoside from blackcurrant seeds along with several anthocyanins and flavonoids.

The presence of phenolic acid glycosides has also been reported in the genus *Fragaria*, with several hexoses of coumaric, ferulic, caffeic and sinapic acids reported primarily in the strawberry receptacle and in small amounts in the achenes (Fait et al., 2008). These differences in the diversity of compounds and their concentrations could be attributed to differences among species, cvs, extraction and detection methods.

3.1.3. Characterization and quantification of anthocyanins

A total of eight anthocyanins were detected in the pink and red gooseberry varieties. Cyanidin-3-rutinoside and cyanidin-3-glucoside were identified by comparison with authentic standards. The other anthocyanins were tentatively identified based on their accurate mass and the fragmentation pattern acquired with MS/MS in different voltages. (Table 3). The major daughter ion produced was the aglycon after the loss of the sugar moiety. Peak 3 corresponded to peonidin-3-glucoside, with small amounts of cyanidin-3-xyloside, also co-eluting under the same peak as revealed by the mass profile obtained with Q-ToF/MS/MS. Peak 4 was tentatively identified as peonidin-3-rutinoside. Peaks 6 and 7 had accurate mass at *m/z* 595.1447, and 595.1480, respectively, and after fragmentation they both yielded a major daughter ion with *m/z* 287.055 which was attributed to cyanidin aglycon. Peak **7** additionally had a daughter ion with *m/z* 147.0570 which is a characteristic fragment of *p*-coumaric acid. Jordheim et al. (2007), have characterised two isomers of cyanidin-3-*O*-*p*-coumaroyl-glucopyranoside in gooseberries by HPLC-DAD and NMR spectroscopy. Peak 7, furthermore had the same UV/Vis spectra (Fig. S1) as cyanidin-3-*O*-β-(6″-*E*-*p*-coumaroylglucopyranoside) as reported by Jordheim et al. (2007). Therefore it was tentatively identified as cyanidin-*p*-coumaroyl-glucopyranoside. Peak 8 exhibited a similar mass profile as peak 4 and it was tentatively identified as peonidin-3-glycoside. Peak 5 was present in very low amounts and it was not possible to acquire a clear LC/MS profile.

The anthocyanin concentrations in pigmented gooseberry cvs are presented in Table 5. Cyanidin-3-rutinoside (pigment 2) was the predominant anthocyanin identified in all the cvs followed by cyanidin-3-glucoside (pigment 1) and pigment 7. 'Scotch Red Rough' exhibited the highest total anthocyanin content with a sum of 265.8 mg kg^{-1} (FW), which is comparable with the levels of anthocyanins reported for other red gooseberry cvs and similar to the levels found in red currants (Djordjevic et al., 2010, Jordheim et al., 2007 and Määttä-Riihinen et al., 2004).

Table 5. Concentration of anthocyanins (mg kg^{-1}) in 20 gooseberry cvs on a FW and DW basis (values inside brackets). The results are presented in descending order according to the sum of anthocyanins per FW.[a]

Cultivar	1		2		3		4		5		6		7		8	
Scotch Red Rough	48.3^{d}	(267.5)	128.8^{h}	(714.6)	3.3^{c}	(18.5)	14.5^{c}	(82.9)	6.0b^{a}	(31.9)	5.7^{c}	(31.7)	57.3^{e}	(319.8)	1.9^{a}	(10.9)
May Dulie	33.6^{d}	(239.0)	36.3^{g}	(256.6)	2.0^{b}	(14.0)	2.9^{b}	(20.6)	1.5^{b}	(10.7)	2.6^{b}	(18.2)	21.1^{d}	(149.9)	0.4^{b}	(3.1)
Careless 3024	8.2^{ab}	(61.6)	19.5^{f}	(146.4)	t^{b}		0.8^{a}	(6.2)	t		0.8^{a}	(5.7)	4.6^{c}	(34.4)	t	
Victoria	9.4^{b}	(83.9)	13.5^{c}	(109.7)	nd^{c}		t		t		0.5^{a}	(4.4)	2.8^{a}	(23.1)	nd	
Coutray's Earliest	6.1^{ab}	(53.6)	8.3^{abc}	(72.0)	0.8^{a}	(6.9)	0.9a	(7.7)	0.6c	(5.0)	t		3.0^{a}	(26.1)	t	
Ajax	5.3^{ac}	(36.1)	9.3^{bc}	(63.1)	t		1.1a	(7.7)	t		t		2.1^{a}	(14.2)	nd	
Lord Audley	3.6^{c}	(30.5)	4.8^{a}	(40.9)	t		t		t		nd		1.4^{b}	(11.8)	nd	
Heart of Oak	1.0^{e}	(6.2)	5.4^{ab}	(31.7)	nd		t		nd		t		t		nd	
Bedford Yellow	t		1.8^{e}	(12.2)	t		t		t		t		t		t	
Nailer	t		1.2^{de}	(7.3)	nd		t		nd		nd		t		t	
Jolly Amylers	t		0.6^{d}	(4.3)	nd		t		nd		t		t		nd	
Mean	14.6	(9.73)	20.9	(132.6)	2.0	(13.2)	4.0	(25.0)	2.7	(15.9)	2.4	(15.0)	13.2	(82.8)	1.2	(7.0)
LSD					0.58	(4.05)	1.17	(2.75)	0.38	(2.76)	0.68	(4.22)			0.69	(4.59)

[a] The cultivars for which the levels of anthocyanins were below the detection limit (DL) or quantification limit (QL) have been omitted from the table.
[b] t = traces.
[c] nd = not detected. Means within the same column with no letters in common are significantly different. Data for which LSD is not provided have been log transformed and the significant differences are based on the analysis of variance (ANOVA) for the logarithmic data.

3.2. Presence of non-structural carbohydrates

The concentration of total sugars on a FW basis, ranged between 31.4 g kg^{-1} to 80.2 g kg^{-1} (Table 6), which is consistent with previous reports (Mikulic-Petkovsek et al., 2012b and Viljakainen et al., 2002). Fructose and glucose were the predominant sugars in all cvs accounting for ~49% and ~40.5%, respectively, of total sugars. The mean ratio of sucrose was ~10% of the total sugars in gooseberry cvs. These results are in agreement with Giné Bordonaba and Terry (2008) who reported the same ratios of sugars in 17 UK-grown blackcurrant cvs, indicating a similar sugar profile of different species belonging to the *Grossulariaceae*. The ratio of fructose to glucose reported here varied from 0.9 to 1.3 and is within the range reported in red and blackcurrants (Milivojevic et al., 2012). The relative proportion of individual sugars in fruits can influence the taste properties such as sweetness (Milivojevic et al., 2012).

Table 6. Concentration of non-structural carbohydrates (g kg^{-1}) in 20 gooseberry cultivars on a FW and DW basis (values inside brackets).[a]

Cultivar	Fructose		Glucose		Sucrose		Total	
Guy seedling	37.5^{h}	(189.8)	31.7^{f}	(160.6)	10.9^{e}	(55.0)	80.2^{g}	(405.4)
Jubilee Careless	34.1^{gh}	(228.4)	29.0^{ef}	(193.5)	9.4^{de}	(61.4)	72.5^{fg}	(483.3)
Ajax	33.2^{fgh}	(225.2)	28.2^{def}	(191.7)	6.4^{cd}	(43.5)	67.8^{fg}	(460.3)
Scotch Red Rough	30.9^{fgh}	(173.9)	24.5^{cdef}	(138.2)	11.5^{e}	(63.4)	66.9^{fg}	(375.5)
Careless (Kent)	32.1^{fgh}	(212.9)	28.0^{def}	(185.9)	6.3^{cd}	(41.8)	66.4^{fg}	(440.6)
Rubuste Nool	32.5^{fgh}	(218.2)	27.4^{cdef}	(183.8)	6.0^{cd}	(44.0)	65.9^{fg}	(442.4)
White Eagle	30.8^{fgh}	(196.8)	27.7^{cdef}	(176.3)	6.6^{cd}	(41.9)	65.1^{fg}	(415.0)
Careless VT	31.6^{fgh}	(242.4)	27.3^{cdef}	(209.3)	4.5^{bc}	(34.8)	63.5^{fg}	(486.4)
Bedford Yellow	31.0^{fgh}	(218.2)	25.7^{cdef}	(180.7)	5.7^{c}	(40.0)	62.4^{fg}	(439.0)
Nailer	27.8^{efgh}	(166.1)	22.1^{bcde}	(132.1)	10.8^{e}	(64.0)	60.7^{f}	(362.2)
Heart of Oak	24.7^{cdef}	(143.5)	20.3^{bcd}	(118.1)	14.3^{e}	(82.5)	59.3^{f}	(344.2)
Cousen's Seedling	29.3^{efgh}	(235.4)	24.9^{cdef}	(200.2)	3.1^{ab}	(25.0)	57.4^{ef}	(460.7)
Mitre	27.3^{efg}	(195.0)	23.3^{cdef}	(166.4)	5.3^{c}	(37.7)	55.9^{def}	(399.1)
May Dulle	26.1^{defg}	(185.7)	22.0^{bcde}	(156.3)	4.6^{c}	(32.8)	52.7^{cdef}	(374.8)
Jolly Amylers	25.0^{bcdef}	(159.4)	20.7^{bc}	(131.9)	5.5^{c}	(35.2)	51.3^{bcdef}	(326.6)
Victoria	21.9^{abcde}	(178.8)	16.4^{ab}	(133.9)	2.9^{a}	(23.5)	41.2^{abcde}	(336.2)
Lord Elco	19.5^{abcd}	(155.7)	16.3^{ab}	(129.0)	2.9^{a}	(22.8)	38.6^{abcd}	(308.5)
Goutray's Earliest	19.3^{abc}	(168.5)	14.2^{a}	(123.4)	3.1^{a}	(27.2)	36.6^{abc}	(319.0)
Lord Audley	17.8^{ab}	(153.0)	14.0^{a}	(118.4)	2.8^{a}	(23.9)	34.6^{ab}	(292.6)
Careless 3024	16.6^{a}	(124.1)	12.0^{a}	(90.0)	2.8^{a}	(21.2)	31.4^{a}	(235.3)
Mean	27.5	(188.4)	22.8	(156.0)	6.3	(40.9)	56.5	(385.4)

[a] The results are presented in descending order according to total sugar content for FW. Means within the same column with no letters in common are significantly different. Results are based on the analysis of variance (ANOVA) for the log transformed data for all the sugars.

3.3. Total ascorbic acid

Ascorbic acid was determined as total ascorbic acid which comprised ascorbic acid and its oxidised form (dehydroascorbic acid). It was detected as a single peak at 3.1 min. The vitamin C content varied from 0.38 g kg^{-1} in 'White Eagle' to 0.85 g kg^{-1} (FW) in 'Ajax' (Table 7). These concentrations are up to 3-fold higher than previous reported for gooseberries (Häkkinen et al., 1999b, Pantelidis et al., 2007 and USDA, 2014), although they are similar to the concentrations reported in blackcurrants (Djordjevic et al., 2010), red currant, strawberry and lingonberry (Skrede et al., 2012) and within the range found in common berries (<1 g kg^{-1} FW) (Giné Bordonaba and Terry, 2011). The differences between these and earlier studies could reflect differences in sample preparation and analytical methods or more importantly differences among varieties.

3.4. Postharvest treatment of gooseberries

The two cvs selected for the storage trial in the following year, included the red cv 'Scotch Red Rough' which exhibited the most diverse profile of phenolics and had a high concentration of anthocyanins as well as the green cv 'Careless (Kent)' which is a popular commercial variety of

Table 7. Concentration of total ascorbic acid (g kg^{-1}) in 20 gooseberry cvs on a FW basis.[a]

Cultivar	Total ascorbic acid
Ajax	0.85
Careless VT	0.83
Jubilee Careless	0.81
Lord Elco	0.74
Careless (Kent)	0.74
Nailer	0.68
Scotch red rough	0.64
Heart of Oak	0.60
Victoria	0.56
Bedford yellow	0.55
Guy Seedling	0.54
Mitre	0.53
Cousen's Seedling	0.53
Careless 3024	0.52
May Dulle	0.51
Lord Audley	0.48
Robuste Nool	0.45
Goutrey Earliest	0.44
Jolly Amylers	0.43
White Eagle	0.38
Mean	0.59
LSD	0.049

[a] The results are presented in descending order according to total ascorbic acid content.

gooseberry. For both cvs (per DW), the phenolic and sugar content at harvest point was compared between the two successive years. The composition of non-structural sugars between the two seasons was similar, with only sucrose in 'Careless Kent' dropping from 41.8 g kg^{-1} in the first year to 26.5 g kg^{-1} in the second ($p < 0.05$) (Table S1). Overall, the profile of phenolics was similar for both years; the most notable difference was the 2-fold drop in anthocyanin content in 'Scotch Red

Rough' during the second year from 1477.8 mg kg^{-1} to 749.5 mg kg^{-1} (Table S1).

3.4.1. Biochemical and physiology changes during storage

In the present study the changes in major phenolic compounds, non-structural sugars, colour and ethylene emission were recorded for both cvs, in order to determine the biochemical and physiology changes in gooseberry fruit during air storage and the effect of external ethylene exposure. The water content was also monitored for both cvs throughout the storage period. 'Careless Kent' had a consistently higher water content compared to 'Scotch Red Rough' with average values at baseline of 87.7% and 81.8%, respectively. The results indicated a marginal water loss of ≤2% for both cvs, irrespective of treatment, which occurred after 7 days of storage (Table S2). Also there was no leakage recorded in any of the cvs.

3.4.2. Ethylene

The change in ethylene production rates for treated and control samples is presented inFig. 2. Ethylene emission remained at low levels throughout the storage period with a total mean of 5.8×10^{-3} ng kg^{-1} s^{-1}, which is significantly lower compared to other soft fruits such as blackberries and raspberries (0.035–0.35 ng kg^{-1} s^{-1} at 5 °C) (Mitcham et al., http://postharvest.ucdavis.edu/PFfruits/Bushberries/). An increasing trend was observed for the red cv with the control samples showing a sharp increase towards the middle of the storage period, but subsequently fell in the same levels as the treated samples. Ethylene emission levels remained relatively stable for the green variety throughout the storage period.

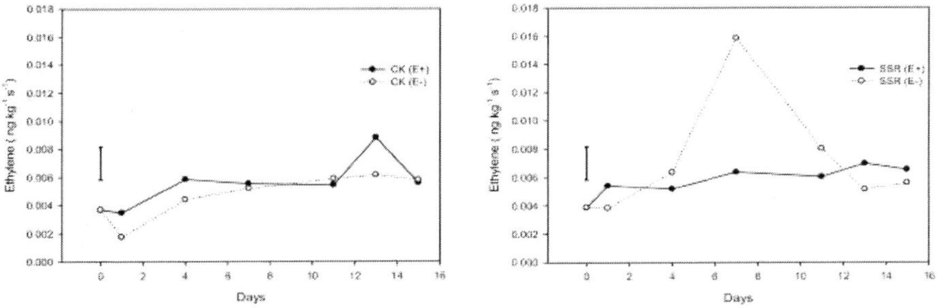

Figure 2. Ethylene emission recorded during the storage period for the two gooseberry cvs, (E +) = ethylene treated samples, E(−) = control samples, CK = 'Careless (Kent)', SSR = 'Scotch Red Rough'. The error bars represent the least significant difference (LSD) at 95% confidence level.

3.4.3. Phenolic compounds

The biochemical results, on a dry weight basis, showed that flavonol glycosides, which represent the major group of phenolic compounds in gooseberries, remained relatively stable for both treated and control samples throughout the storage period. 'Scotch Red Rough' had consistently higher concentrations of flavonol glycosides during the trial, as it is characteristic of the cv (Table S3).

Anthocyanins on the other hand exhibited an increasing trend in 'Scotch Red Rough' over time irrespective of treatment (Fig. 3). A marginal increase observed for the ethylene treated samples was not recorded towards the end of the storage period with both treated and control samples, reaching the same anthocyanin concentrations. This could indicate an initial positive effect of exogenous ethylene on the accumulation of anthocyanins which was counteracted by the sudden increase in ethylene production observed in the control samples towards the middle of the storage period. Other non-climacteric fruits, such as grapes, have exhibited a clear response to postharvest ethylene treatment with anthocyanins positively affected at the end of the treatment period (Becatti et al., 2010 and Becatti et al., 2014). These differences could be due to species differences and also differences in the treatment protocols applied. Grape berries were exposed to very high ethylene concentrations (1000 mg L^{-1}) and for a longer period of time (36 h), which could explain the acute metabolic responses to ethylene.

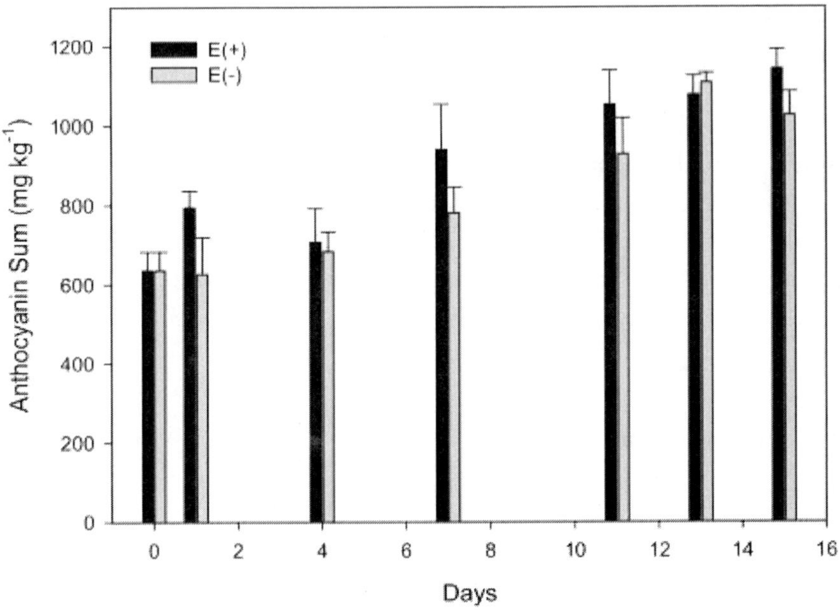

Figure 3. Concentration of total anthocyanins on a DW basis, in 'Scotch Red Rough' during the storage period. E(+) = Ethylene treated samples, E(−) = control. The error bars represent the standard error (SE) of the mean.

On the other hand, the accumulation of anthocyanins under air-storage, observed for the red gooseberry cv, seems to be a common trend among berry species, Kim et al. (2015), recently have demonstrated an increase in anthocyanin content in blackberries during storage at 1 °C, which was more pronounced if the berries were transferred to 20 °C towards the end of the storage period.

Earlier reports, suggest that gooseberry exposure to ethylene, can cause some yellowing/browning of the skin (Cantwell, 2002). In the current study, colour measurements revealed a decrease in L^* and C^* values over time, for both treated and control gooseberries (Table S2), which was not associated with any perceived colour deterioration. The decrease in L^* values could be attributed to several reasons including the observed accumulation of anthocyanin and perhaps gradual water loss. The change in C^* values, early in the storage period, may be related to an initial response to exogenous ethylene for the treated samples and the increase in ethylene emission for the control samples, which was further enhanced by water loss later in the storage period.

3.4.4. Non-structural carbohydrates

The concentrations of non-structural carbohydrates, for both cvs examined, remained relatively stable for both treated and control samples. The main difference in sugar composition between the two cvs was in the levels of sucrose, which were particularly elevated in 'Scotch Red Rough' throughout the storage period. The average concentration of total sugars for the two cvs, at baseline and during the storage period, is shown in Fig. 4. A previous study on 'Achilles' gooseberries stored under air and different CA conditions, showed that total soluble solids, remained stable for 7 weeks (Harb and Streif, 2004).

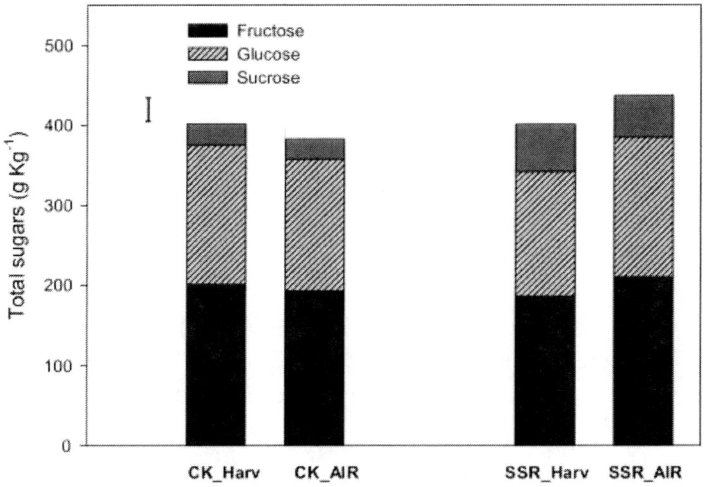

Figure 4. Average concentrations of non-structural carbohydrates on a DW basis, before and during storage for the two gooseberry cvs. CK_Harv = 'Careless (Kent)' at harvest, CK_AIR = 'Careless (Kent)' under air storage, SSR_Harv = 'Scotch Red Rough' at harvest, SSR_AIR = 'Scotch Red Rough' under air storage.

The biochemistry and physiology results overall, reveal that gooseberries did not show any clear response to the application of exogenous ethylene and are in broad agreement with previous reports classifying gooseberries as fruits with low sensitivity to ethylene (Cantwell, 2002).

4. CONCLUSIONS

The present study reveals the potential for the use of gooseberries as a rich source of bioactive compounds and demonstrates the distinguishing variability across cultivars and tissue types. Furthermore, it enhances our understanding of their postharvest behaviour especially in relation to their response to exogenous ethylene and the preservation of their nutritional quality under cold storage.

ACKNOWLEDGEMENTS

This study was supported by Unilever U.K. Central Resources Ltd, TSB 101125 (Innovate UK), and the Biotechnology and Biological Sciences Research Council (BBSRC). The authors are grateful to Dr M. Ordidge from the University of Reading for providing access to the National Fruit Collection and Farm Advisory Services Team (FAST) for helping with the sample collection.

REFERENCES

1. Bakowska-Barczak, A.M., Schieber, A., Kolodziejczyk, P., 2009. Characterization of Canadian Black Currant (Ribes nigrum L.) Seed Oils and Residues. J. Agric. Food Chem. 57, 11528–11536.

2. Barney, D.L., Hummer, K.E., 2005. Currants, Gooseberries, and Jostaberries. A Guide for Growers, Marketers, and Researchers in North America. The Haworth Press, Binghamton, NY, USA, pp. 266.

3. Becatti, E., Ranieri, A., Chkaiban, L., Tonutti, P., 2010. Ethylene and wine grape berries: metabolic responses following a short-term postharvest treatment. Acta Hortic. 223–228.

4. Becatti, E., Genova, G., Ranieri, A., Tonutti, P., 2014. Postharvest treatments with ethylene on Vitis vinifera (cv Sangiovese) grapes affect berry metabolism and wine composition. Food Chem. 159, 257–266.

5. Cantwell, M., 2002. Summary Table of Optimal Handling Conditions for Fresh Produce. In: Kader, A. (Ed.), Postharvest Technology of Horticultural Crops, 3. UCANR Publications, Oakland, California, USA, pp. 511–518.

6. Chiang, C., Kadouh, H., Zhou, K., 2013. Phenolic compounds and antioxidant properties of gooseberry as affected by in vitro digestion. LWT – Food Sci. Technol. 51, 417–422.

7. Dale, A., 2000. Potential for Ribes cultivation in North America. Hort. Technol. 10, 548–554.

8. Djordjevic, B., Šavikin, K., Zdunic, G., Jankovic, T., Vulic, T., Oparnica, C., Radivojevic, D., 2010. Biochemical properties of red currant varieties in relation to storage. Plant Foods Hum. Nutr. 65, 326–332.

9. Fait, A., Hanhineva, K., Beleggia, R., Dai, N., Rogachev, I., Nikiforova, V.J., Fernie Aharoni, A.R.A., 2008. Reconfiguration of the achene and receptacle metabolic networks during strawberry fruit development. Plant Phys. 148, 730–750.

10. Filipiak-Szok, A., Kurzawa, M., Szlyk, E., 2012. Determination of anti-oxidant capacity and content of phenols, phenolic acids, and flavonols in Indian and European gooseberry. Chem. Pap. 66, 259–268.

11. Folmer, F., Basavaraju, U., Jaspars, M., Hold, G., El-Omar, E., Dicato, M., Diederich, M., 2014. Anticancer effects of bioactive berry compounds. Phytochem. Rev. 295– 322.

12. Giné Bordonaba, J., Terry, L.A., 2008. Biochemical profiling and chemometric analysis of seventeen UK-grown black currant cultivars. J. Agric. Food Chem. 56, 7422–7430.

13. Giné Bordonaba, J., Terry, L.A., 2011. Ribes and Rubus [blackberry, currants and raspberry, etc.]. In: Terry, L.A. (Ed.), Health-Promoting Properties of Fruits and Vegetables, 1. Cabi, Oxford, UK, pp. 260–290.

14. Gopevac, D., Teševi⍰c, V., Vajs, V., Milosavljevi⍰c, S., Zduni⍰c, G., Orpevii c, B., Stankovii c, M., 2012. Chemical composition of currant seed extracts and their protective effect on human lymphocytes DNA. J. Food Sci. 77, C779–C783.

15. Häkkinen, S., Heinonen, M., Kärenlampi, S., Mykkänen, H., Ruuskanen, J., Törrönen, R.,1999a. Screening of selected flavonoids and phenolic acids in 19 berries. Food Res. Int. 32, 345–353.

16. Häkkinen, S.H., Kärenlampi, S.O., Heinonen, I.M., Mykkänen, H.M., Törronen, A.R., 1999b. Content of the flavonols quercetin, myricetin, and kaempferol in 25 edible berries. J. Agric. Food Chem. 47, 2274–2279.

17. Harb, J., Streif, J., 2004. Quality and consumer acceptability of gooseberry fruits (Ribes uva crispa) following CA and air storage. J. Hortic. Sci. Biotech. 79, 329– 334.

18. Hummer, K.E., Dale, A., 2010. Horticulture of Ribes. For. Path. 40, 251–263.

19. Jordheim, M., Måge, F., Andersen, Ø.M., 2007. Anthocyanins in berries of Ribes including gooseberry cultivars with a high content of acylated pigments. J. Agric. Food Chem. 55, 5529–5535.

20. Kampuse, S., Muizniece-Brasava, S., Kruma, Z., Sabovics, M., Savri, S., Ruse, K., 2015. The influence of pretreatment technology and packaging material on the quality of fresh gooseberries during storage. Acta Hortic. 1071, 405–412.

21. Kim, M.J., Perkins-Veazie, P., Ma, G., Gina Fernandez, G., 2015. Shelf life and changes in phenolic compounds of organically grown blackberries during refrigerated storage. Postharvest Biol. Technol. 110, 257–263.

22. Lu, Y., Foo, L.Y., 2003. Polyphenolic constituents of blackcurrant seed residue. Food Chem. 80, 71–76.

23. Lu, Y., Foo, L.Y., Herbert Wong, H., 2002. Nigrumin-5-p-coumarate and nigrumin-5- ferulate, two unusual nitrile-containing metabolites from black currant (Ribes nigrum) seed. Phytochemistry 59, 465–468.

24. Määttä-Riihinen, K.R., Kamal-Eldin, A., Mattila, P.H., González-Paramás, A.M., Törrönen, R., 2004. Distribution and contents of phenolic compounds in eighteen scandinavian berry species. J. Agric. Food Chem. 52, 4477–4486.

25. Mattila, P., Hellström, J., Törrönen, R., 2006. Phenolic acids in berries, fruits, and beverages. J. Agric. Food Chem. 54, 7193–7199.

26. McKay, S., Van Eck, A., 2006. Red currants and gooseberries: extended season and marketing flexibility with controlled atmosphere storage. N. Y. Fruit Q. 14, 43– 45.

27. Mikulic-Petkovsek, M., Slatnar, A., Stampar, F., Veberic, R., 2012a. HPLC-MSn identification and quantification of flavonol glycosides in 28 wild and cultivated berry species. Food Chem. 135, 2138–2146.

28. Mikulic-Petkovsek, M., Schmitzer, V., Slatnar, A., Stampar, F., Veberic, R., 2012b. Composition of sugars, organic acids, and total phenolics in 25 wild or cultivated berry species. J. Food Sci. 77, C1064–C1070.

29. Milivojevic, J., Slatnar, A., Mikulic-Petkovsek, M., Stampar, F., Nikolic, M., Veberic, R., 2012. The influence of early yield on the accumulation of major taste and health-related compounds in black and red currant cultivars (Ribes spp.). J. Agric. Food Chem. 60, 2682–2691.

30. Mitcham, E., Crisosto, C.H., Kader, A.A., 2015. Bushberries: Blackberry, Blueberry, Cranberry, Raspberry: Recommendations for Maintaining Postharvest Quality. . (accessed 20-10-2015) http://postharvest.ucdavis.edu/PFfruits/Bushberries/. Mitchell, C., Brennan, R.M., Cross, J.V., Johnson, S.N., 2011. Arthropod pests of currant and gooseberry crops in the U.K.: their biology, management and future prospects. Agric. For. Entomol. 13, 221–237.

31. Muizniece-Brasava, S., Dukalska, L., Kampuse, S., Murniece, I., Savri, S., DabinaBicka, I., 2015. The evaluation of physical parameters of gooseberries using different pretreatment technologies and packaging materials. Acta Hortic. 1071, 381–388.

32. Pantelidis, G.E., Vasilakakis, M., Manganaris, G.A., Diamantidis, G., 2007. Antioxidant capacity, phenol, anthocyanin and ascorbic acid contents in raspberries, blackberries, red currants, gooseberries and cornelian cherries. Food Chem.102, 777–783.

33. Planchon, V., Lateur, M., Dupont, P., Lognay, G., 2004. Ascorbic acid level of Belgian apple genetic resources. Sci. Hortic. 100, 51–61.

34. Pluta, S., Broniarek-Niemiec, A., Zurawicz, E., 2010. Productive value of eighteen gooseberry (Ribes grossularia L.) cultivars of different origin evaluated in central Poland. J. Fruit Ornam. Plant Res. 18, 197–209. Russell, W.R., Labat, A., Scobbie, L., Duncan, G.J., Duthie, G.G., 2009.

35. Phenolic acid content of fruits commonly consumed and locally produced in Scotland. Food Chem. 115, 100–104.

36. Skrede, G., Martinsen, B.K., Wold, A., Birkeland, S., Aaby, K., 2012. Variation in quality parameters between and within 14 Nordic tree fruit and berry species. Acta Agric. Scand. Sec. B Soil Plant Sci. 62, 193–208.

37. Terry, L.A., Chope, G.A., Giné Bordonaba, J., 2007. Effect of water deficit irrigation and inoculation with Botrytis cinerea on

strawberry (Fragaria x ananassa) fruit quality. J. Agric. Food Chem. 55, 10812–10819.

38. Terry, L.A., Crisosto, C.H., Forney, C.F., 2009. Small fruit and berries. In: Yahia, E.M. (Ed.), Modified and Controlled Atmospheres for the Storage, Transportation, and Packaging of Horticultural Commodities. CRC Press, Boca Raton, USA, pp. 363– 395.

39. USDA, 2014. National Nutrient Database for Standard Reference Release 27. . (accessed 05-02-2015) http://ndb.nal.usda.gov/ndb/foods/show/2268?fg=Fruits+and+Fruit+Juices&man=&lfacet=&format=Abridged&count=&max=25&offset=75&sort=&qlookup=.

40. Thompson, J.F., 2002. Transportation. In: Kader, A. (Ed.), Postharvest Technology of Horticultural Crops, 3. UCANR Publications, California, USA, pp. 259–269.

41. Viljakainen, S., Visti, A., Laakso, S., 2002. Concentrations of organic acids and soluble sugars in juices from nordic berries. Acta Agric. 52, 101–109.

42. Wang, L.-S., Stoner, G.D., 2008. Anthocyanins and their role in cancer prevention. Cancer Lett. 269, 281–290.

43. Wu, X., Gu, L., Prior, R.L., McKay, S., 2004. Characterization of anthocyanins and proanthocyanidins in some cultivars of Ribes,Aronia, and Sambucus and their antioxidant capacity. J. Agric. Food Chem. 52, 7846–7856.

11

CONTROLLED WATER STRESS TO IMPROVE FRUIT AND VEGETABLE POSTHARVEST QUALITY

Leonardo Nora, Gabriel O. Dalmazo

Fabiana R. Nora and Cesar V. Rombaldi University Federal of Pelotas Brazil

1. INTRODUCTION

Healthier food produced in a sustainable manner at affordable price is a necessity in the contemporaneous society. In this context, studies to devise the minimum necessary amount of water to crops towards development are required not only to save water and/or energy, but also to improve plant fitness to cope with biotic and abiotic stresses, even after harvest, and to improve nutritional, functional and sensorial food properties. The awareness of the growing impact of environmental stress has lead to worldwide efforts in adapting agricultural production to adverse environmental conditions, focusing on mitigating quantitative yield losses (Godfray et al., 2010). Far less attention has been devoted to the impact of abiotic environmental stresses on crop quality (Wang & Frei, 2011). Limited water supply (LWS) is a serious threat to agriculture, and often cause yield reduction. However, many plant species are not intolerant to that, otherwise the water input concept should be revised. In this review we describe the effect of limiting water supply during plant development on fruit and vegetables postharvest quality, mainly in terms of sensorial attributes (texture, colour, aroma, and taste) and composition (nutrients and bioactive compounds), and also in terms of gene expression, enzyme activity, yield, and storage behaviour. A brief introduction about regulated deficit irrigation, plant growth regulators and secondary metabolites is presented to illustrate common aspects associated with water stress effects on fruit and vegetable quality.

2. REGULATED DEFICIT IRRIGATION

Deficit irrigation (i.e. irrigation below optimal crop water requirements) research to improve productivity of horticultural crops began in the 1970's with the aim to control excessive vegetative vigour in high-density orchards. Tree physiology was intensively studied to examine timing of water deficits that would minimize the impact on fruit growth but maximize effects on shoot growth. This strategy later became known as regulated deficit irrigation (RDI). In recent years there has been resurgence in the application of RDI to horticultural crops due to changes in climate resulting in severe drought. However, focus has now switched from controlling excessive vigour to investigating opportunities to stimulate improvements in fruit and vegetable quality so that any yield loss can be compensated by an increase in crop value (Costa et al., 2007; Stefanelli et al., 2010).

3. PLANT GROWTH REGULATORS

Organisms need to adapt themselves to changes in fluctuating environmental conditions. The plants, since they are not able to scape from adverse environmental conditions, have to rely entirely on their developmental plasticity to survive (Krouk et al., 2011). These adaptations include the responses to temperature fluctuations, water and nutrients imbalance, UV radiation, pathogens, and insects, among other biotic and abiotic stresses. Plant growth regulators (phytohormones), compounds derived from plant biosynthetic pathways, mediate these responses by acting either at the site of synthesis or following their transport, elsewhere in the plant. Collectively, plant hormones regulate every aspect of plant growth, development and the responses of plants to biotic and abiotic stresses (Peleg & Blumwald, 2011). Classical phytohormones are abscisic acid (ABA), ethylene, cytokinin (CK), auxin, gibberellin, jasmonate, as well as brassinosteroids, salicylic acid, nitric oxide, and strigolactone, and it is likely that additional growth regulators are yet to be discovered (Santner & Estelle, 2009). ABA synthesis is one of the fastest responses of plants to water stress, triggering ABA-inducible gene expression and causing stomatal closure, thereby reducing water loss via transpiration and eventually restricting cellular growth (Wilkinson & Davies, 2010; Yamaguchi-Shinozaki & Shinozaki, 2006). Many ABA-mediated physiological processes induced by water deficit, including closure of the stomata and acceleration of leaf senescence, are counteracted by CKs which increase stomatal aperture and/or delay ABA-induced stomatal closure. It has been suggested that in longer-term responses to stress, hormones such as ABA and CK may function to regulate the production, metabolism and distribution of metabolites essential for stress survival and recovery (Pospíšilová & Dodd, 2005; Stoll et al., 2000).

4. SECONDARY METABOLITES

Plants produce a huge variety of secondary metabolites with roles in various biological processes, such as pollination, seed dispersal, and resistance to biotic and abiotic stresses (Wink, 1999). Until recently it was thought that genes for plant metabolic pathways were not clustered, and this is certainly true in many cases. However, five plant secondary metabolic gene clusters have now been discovered, all of them implicated in synthesis of defence compounds, with enzymes for the first committed steps apparently recruited directly or indirectly from primary metabolic pathways involved in hormone synthesis (Chu et al., 2011). The genes and corresponding gene products for the first committed steps in these pathways can be regarded as signature genes/enzymes, as they are required for the synthesis of the skeleton structures of the different classes of secondary metabolite (Osbourn, 2010). The signature genes all share homology with genes from plant primary metabolism, and so are likely to have been recruited directly or indirectly from primary metabolism by gene duplication and acquisition of new functions (Chu et al., 2011). Plant secondary metabolites have long been associated with ecological roles in antagonistic or mutualistic interactions between plants and their herbivores, pathogens, competitors, pollinators or seed dispersers. However, many of these substances have been demonstrated to function in the primary processes of growth and development or resistance to abiotic stresses (Fig. 1). Clearly, more attention should now be devoted to looking for internal roles. In the past decade, several substances that were once considered to be secondary metabolites, such as jasmonic acid, salicylic acid and brassinosteroids, have been shown to be important internal signals (D'Auria & Gershenzon, 2005).

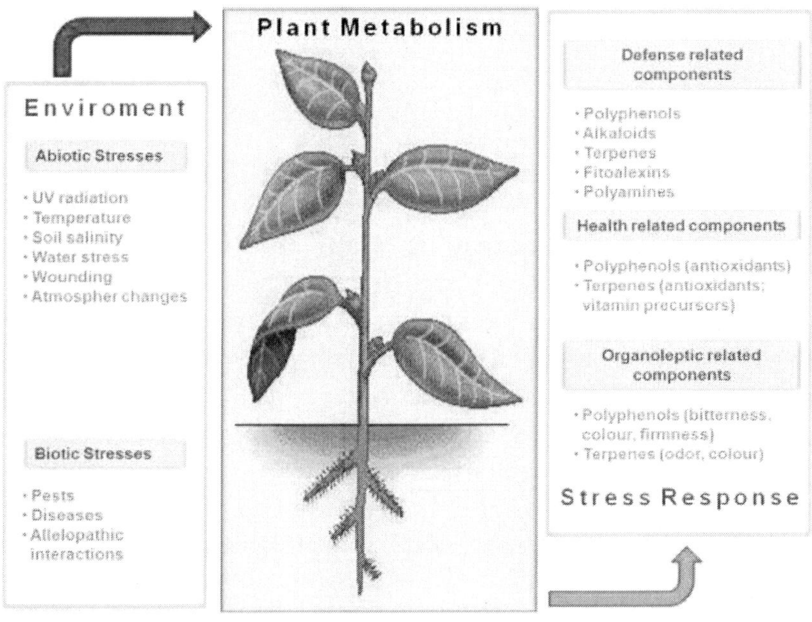

Figure 1. Schematic representation of plant metabolism response to environmental stress.

5. WATER STRESS EFFECTS ON FRUIT QUALITY

5.1 Kiwi

Miller et al. (1998), performed a two year experiment in New Zealand to determine the responses of kiwifruit (Actinidia deliciosa cv. Hayward) to water stress conditions. Plants were submitted to a stress regime composed of three different treatments: Control (water received according to culture demands), water stress in early summer and water stress in late summer. They observed a significant loss in fruit weight, especially in plants exposed to stress in early summer (fruit set period). In contrast, an increase in the total soluble solids occurred. Differences in firmness and performance during storage were not detected for stressed kiwifruit. However, in a similar experiment conducted by Reid et al. (1996), kiwifruit harvested from vines exposed to a less severe drought stress were unaffected in size and fruit firmness was retained for 30 days longer in comparison to control (fruit harvested from fully irrigated vines).

5.2 Strawberry

Bordonaba & Terry (2010), testing strawberry (Fragaria X ananassa Duch. cvs. Elsanta, Sonata, Symphony, Florence and Christine) response to water deficits obtained promising results. Stressed plants received a quarter of irrigation water compared to control plants (cultivated under field capacity). Water suppress started when most fruit from the primary truss were at flower initiation stage. The different cultivars responded in a specific manner to water stress, expressing specific water usage. Berry size was equivalent (Florence and Christine) or smaller (Sonata and Symphony) than control plants. Dry mater, as proportion of fruit weight, was considerably greater in fruit from water stressed plants than from plants kept at or near field capacity. Berries from stressed plants showed lower redness (higher h^{o} value) than control berries. Considering that the main components of red colour in strawberries are anthocianins is plausible to presume that these fruit have lower contents of this secondary metabolite. However, in a previous study, Terry et al. (2007) observed the same reduction in red colour, but anthocianins measurements pointed to a higher content of this metabolite. Authors attribute that to an artefact of the objective colorimeter due to smaller berry size. No significant differences were found in sugar contents among treatments. However, monosaccharides (fructose and glucose) were in higher concentrations in stressed plants, thus berries were sweeter. Acids are also important flavor components in strawberries. The stressing condition increased the acidity in all cultivars except Elsanta and Sonata.

5.3 Apple

Mpelasoka et al. (2001b) demonstrated that deficit irrigation (DI) has effects on fruit maturation and ripening depending on timing of application. They tested early deficit irrigation (EDI), applied from 63 to 118 days after full bloom (DAFB), late deficit irrigation (LDI), applied from 118 DAFB to final harvest on 201 DAFB, whole-season deficit irrigation (WDI), irrigated only twice during the late growing season, when volumetric soil water content (θ) declined below 0.15 m3 m-3. These DI treatments all reduced volumetric soil water content. Control consisted in commercially irrigated (CI) trees, irrigated to maintain soil moisture at or close to field capacity. All DI treatments increased fruit total soluble solids (TSS) and firmness regardless of maturity but had little or no effect on titratable acidity. According to the authors the DI fruit may be harvested over a longer period due to their earlier increased TSS and their higher firmness prior to harvest and for most of the storage period. However, the advanced ripening of the DI fruit is responsible for the loss of advantage by DI regarding firmness after long-term storage. Fruit thinning has been proposed as a feasible strategy to compensate the loss in fruit size caused by water stress (Mpelasoka et al., 2001a). Irrigation treatments did not affect the crop load. Irrespective of fruit thinning treatment, deficit irrigated stress resulted in lower fruit weight, total yield and fresh-market yield at harvest than control. However, under deficit treatment, thinned trees resulted higher fruit weight and equal fresh-market yield. Regarding quality parameters, deficit irrigated plants exhibited higher contents of TSS than fully irrigated plants. This could have a positive impact on fruit taste. An equal increase was observed in fruit firmness. Although this attribute correlates to a smaller fruit size due to dehydration during water restriction. In a similar experiment, with apple, the fruit firmness was higher under water restriction treatments compared to fully irrigated treatments despite of fruit size (Mpelasoka et al., 2000). Regarding postharvest conditions, fruit exposed to water restrictions had lower weight loss during cold storage than those originated from fully irrigated treatments. According to the authors, the reduced weight loss can be explained by the structure and/or composition of the skin or the epicuticular waxes covering the skin. This reduction in weight loss could prolong the cold storage life.

5.4 Pear

Lopez et al. (2011) submitted pear (Pyrus communis L. cv. Conference) to water restrictions and fruit thinning at the Stage II (80 and 67 days before harvest in 2008 and 2009, respectively). The experiment was composed of two simple irrigation treatments: fully irrigated plants (100% Evapotranspiration, ETc) and deficit irrigated (20% ETc, preceded by three weeks of total water deprivation to induce the stress response). Each irrigation treatment was spliced in two thinning treatments: no thinning (approximately 180 fruit per tree) and thinning (approximately 85 fruit per tree). Fruit thinning has been proposed as a feasible strategy to compensate the loss in fruit size caused by water stress (Mpelasoka et al., 2001c). Different Irrigation regimes did not affect the crop load between irrigation treatments. Irrespective of fruit thinning treatment, deficit irrigated trees had lower weight, total yield and fresh-market yield at harvest than control. However, under deficit treatment, thinned trees had higher fruit weight and equal fresh-market yield. Regarding quality parameters, deficit irrigated plants exhibited higher contents of total soluble solids than fully irrigated plants. Deficit irrigatedthinned trees had equivalent fruit weights at harvest compared to full irrigated-non thinned trees, confirming that fruit size can be improved by fruit thinning. Considering that fruit size is an important attribute in pear, deficit irrigation can improve fruit marketability.

5.5 Apricot

Perez-Pastor (2007) evaluated postharvest fruit quality of apricot (Prunus armeniaca L. cv. Búlida) harvested from trees exposed to three different treatments: control treatment (100% of evapotranspiration); regulated deficit irrigation (RDI), which consists in fully irrigation during critical periods; and 50% water regime compared to control. At harvest was not observed differences in weight, equatorial diameter and firmness of the fruit among the different treatments. In addition, fruit from water stressed plants had higher values of total soluble solids (TSS), titratable acidity (TA) and hº value (skin colour). During storage, stressed plants had lower decreases in fruit colour (skin and pulp) compared to fully irrigated plants. During the first 20 days of storage, stressed plants conserved higher values of TSS and TA, after that the differences disappeared between treatments. During a simulated retail sale, fungi of the genera Rhizopus, Monilinia, Penicillium, Alternaria, Botrytis and Cladosporium caused fruit losses. Interestingly, a lower fungal attack was observed in stressed fruit. The authors link this fact to a thicker cuticle and/or to the absence of microcrackings. Perez Sarmiento et al. (2010) using nine year-old apricot-trees (Prunus armeniaca L. cv. 'Búlida') grafted on 'Real Fino' rootstock analyzed the effects of RDI on fruit quality. Two irrigation treatments were established. The first, a control treatment, was irrigated to fully satisfy the crop water requirements (100% ETc) during the critical periods (stage III of fruit growth and two months after harvest period), and the second, a RDI treatment, was subject to water shortage during the non-critical periods of crop development, by reducing the amount of applied irrigation water to: a) 40% of ETc from flowering until the end of the first stage of fruit growth; b) 60% of ETc during the second stage of fruit growth and c) 50% and 25% of ETc during the late postharvest period (that starts 60 days after harvesting), for the first 30 days and until the end of tree defoliation, respectively. They found that some qualitative characteristics such as the level of soluble solids, fruit taste and the colour of the fruit are enhanced.

5.6 Peach

Gelly et al. (2003) evaluated the effects of water deficit on fruit quality of peaches (Prunus persica L.). It was shown an increase in soluble solids content and coloration of fruit when during production was applied RDI. Water deficit was applied during either stage II of fruit development (RDI-SII) or during stage II and postharvest (RDI-SII-PH), as compared with non-droughted (control) and postharvest (RDI-PH) treatments. Significant higher concentration of soluble solids and accentuated red colour were observed at harvest in fruit from RDI treatments when compared to control fruit. Accordingly, fruit submitted to RDI during development had greater ethylene production when detached. Ethylene production by RDI-PH fruit did not change, but their quality did in terms of increased soluble solids concentration and improved skin colour, similar to the RDI-SII treatment. Buendía et al. (2008) investigated the influence of regulated deficit irrigation (RDI) on the content of vitamin C, phenolic compounds and carotenoids of peaches. Fruit were harvested from five-year-old early ripening peach trees (cv. "Flordastar", grafted on GF677 rootstock) in Santomera, Murcia, Spain. Two irrigation strategies, fully irrigated (FI) and RDI, were compared at two levels of thinning, commercial and half of the commercial crop load. RDI caused fruit peel stress lowering the content of vitamin C and carotenoids, while increasing the phenolic content, mainly anthocyanins and procyanidins. Fruit weight was the only quality index influenced by the crop load as it increased in FI fruit at low crop load. According to the authors, the decrease in the antioxidant constituents could be due to a higher sunlight exposure of fruit collected from RDI trees as a result of a low vegetative growth of those trees. The increase in the anthocyanin content could be also explained as a response mechanism of peaches against UV irradiation, to mitigate the photooxidative injury of plant tissue.

5.7 Plum

The effects of RDI (RDI) and crop load on Japanese plum (Prunus salicina) cv. Black-Gold were investigated by Intrigliolo & Castel (2010). RDI was applied during phase II of fruit growth and postharvest and compared to a control irrigation treatment (full crop evapotranspiration). Plants from each irrigation treatment were thinned to reach a commercial crop load (described as medium) and to approximately 40% less than the commercial practice (described as low). RDI strategy increased the efficiency of water usage, with 30% of water savings, having minimal effect on crop yield and fruit growth. Economic return, calculated from fruit weight distribution by commercial categories, was more affected by RDI than yield. The combination of medium crop load and RDI shifted fruit mass distribution towards the low value categories. This leads to similar or even higher economic returns in the RDI treatment with low crop level than with the medium one. In addition, since both, low crop level and RDI, increased fruit total soluble solids (TSS), fruit produced under RDI and low crop levels had the highest values of TSS. Deficit irrigation improved fruit composition and, in the short-term, increased tree water use efficiency.

5.8 Citrus

Recently García-Tejero (2010) determined the postharvest fruit quality of oranges (Citrus sinensis L. Osbeck, cv. Salustiano) exposed to RDI in commercial orchards at the semi-arid region of Andalusia- Spain, in the years 2005, 2006 and 2007. The experiment was composed by four different treatments: Control (irrigation replacing 100% of Evapotranspiration, ETc), low deficit irrigation (75% of ETc), moderate deficit irrigation (65% of ETc) and severe deficit irrigation (50% of ETc). As a result, fruit quality parameters as TSS and TA increased in all stressed treatments resulting better organoleptic parameters. Significant fruit size reduction was observed in the year 2005 only. However, reduction in fruit size is often associated with an increase in fruit number (Treeby et al., 2007). In the years 2006 and 2007, a significant yield loss was not observed. The low deficit irrigation treatment proportionate a water saving of 93 mm, which consists in lower irrigation costs that might compensate the slightly yield losses in fruit production. In a similar study Velez et al. (2007), using the "maximum daily trunk shrinkage" method, which is used as an indicator of water stress, was able to obtain water savings reaching 18% without significant decreases in average fruit yield, weight and number. In addition, fruit submitted to deficit irrigation had significantly higher TSS and similar TA. According to the authors, a higher accumulation of sugars is a result of an active response to water deficit.

5.9 Grape

Working with grape (Vitis vinifera L. cv. Rizamat), table type, Du et al. (2008) performed a two year experiment to investigate the alternate partial root-zone drip irrigation on fruit quality, yield and water use efficiency in the arid region of northwest China. The treatments consisted of a control (both sides of the root zone irrigated); an alternate drip irrigation treatment (both sides of the root zone irrigated, one at a time with half the water used in control) and a fix drip irrigation (only one side of the root zone was irrigated with half the water used in control). As a result, in alternate drip irrigation condition, the photosynthetic rate was similar to control whilst the transpiration rate kept in the same level. The leaf water use efficiency increased indicating that the plants submitted to water stress needed less water to keep hydrated. In stressing condition significant yield losses were not perceived, in addition a higher percentage of edible grapes were achieved, which consequently improve product price. A plausible explanation for such performance relays on changes in balance, between vegetative and reproductive growth, of plants facing limited water supplies, resulting in a higher flow of photoassimilates to berries. Stressed grapes presented higher concentrations of both, ascorbic acid and total soluble solids, and lower titrated acidity, culminating in healthier and sweeter grapes. In a similar study (Santos et al., 2007) compared the effects of partial root zone drying irrigation system (50% ETc irrigating one side at a time) with the conventional deficit irrigation system (50% ETc applied on both sides), full irrigation system (100% ETc applied on both sides) and non irrigated vines. The one-year experiment was performed in mature "Moscatel" grapevines (Vitis vinifera L.) in south Portugal. Plants submitted to partial drying regime showed a decreased vegetative growth, expressed by the smaller values of leaf layer number, percentage of water shoots, shoot weight, pruning weight and total leaf area. In synthesis, plants had a higher control over vegetative growth in face of a stressing condition. The partial drying regime and the conventional deficit irrigation treatments, despite receiving the same quantities of water during the experiment, led to different plant responses. The author argue that in the partial drying regime, roots in the dry side produce chemical signals which restrict plant growth. As a consequence, berry temperature increases in response to augmented solar incidence. In response, the concentrations of anthocyanins, phenols and glycosyl-glucose were higher in water stressed plants culminating in higher quality grapes. Yield loss was significant only in non-irrigated treatment.

5.10 Watermelon

Proietti et al. (2008) determined quality parameters of mini-watermelon (Citrullus lanatus) cv. Ingrid, ungrafted or grafted onto a squash hybrid rootstock, and grown under different irrigation regimes: 1.0, 0.75, and 0.5 of evapotranspiration (ETc) rates. The highest fruit yields were observed at 1.0 ETc and at 0.75 ETc when compared to 0.5 ETc, in grafted plants. Grafting of mini-watermelon under irrigation deficit did not modify crop response to water availability, but increased productivity and induced small positive changes in plant quality and nutritional value.

6. WATER STRESS EFFECTS ON VEGETABLE QUALITY 6.1 POTATO

Bejarano et al. (2000) evaluated the content of glycoalkaloids (GAs, α-solanine and α-chaconine) in drought-tolerant potato (Solanum tuberosum L.) grown in the Bolivian highlands under drought stress. Under drought stress conditions GAs concentration increased an average of 43% and 50% in the improved and control cultivars, respectively, but never above the recommended food safety limit (200 mg.kg-1 fresh tubers). GAs are natural toxins synthesised by plants of the Solanaceae family and are believed to be associated with resistance to certain insects. At least 95% of total GAs are in the form of α-solanine and α-chaconine. Both compounds are heat-stable and therefore are not destroyed by common cooking processes such as boiling or frying (Friedman, 2006).

6.2 Tomato

Sánchez-Rodríguez et al. (2011) obtained insightful results regarding the phenolic metabolism in response to water stress. Phenolic compounds are the most abundant type of secondary metabolites in plants, being frequently associated with beneficial effects in human health (Dixon & Paiva, 1995; Hooper & Cassidy, 2006). Tomato plants (Solanum lycopersicum L.), cultivars Kosaco, Josefina, Katalina, Salomé and Zarina, were submitted to water stress conditions. The experimental design consisted of two watering treatments: Control (100% of field capacity) and moderate water stress (50% of field capacity). Zarina, characterized as a drought tolerant cultivar, presented the best responses. All the cultivars had significant decreases in phenolic compounds when submitted to water stress, except Zarina, which showed an increase, particularly in flavonoids, when cultivated under moderate water stress. This increase was concomitant with a 33% higher DAHPS (EC 4.1.2.15) activity, the key enzyme controlling the carbon flow towards phenolic metabolism, and equivalent increases were observed for other related enzymes. The enzymes that degrade phenols, PPO (EC 1.10.3.2) and GPX (EC 1.11.1.7), declined in activity, 30% and 47%, respectively. Thipyapong et al. (2004) propose that in tomato plants a decrease in PPO activity lowers the H2O2 concentration, reducing lipid peroxidation and improving resistance against water stress.

6.3 Lettuce

Coelho et al. (2005) investigated the yield and bioactive amine content of American lettuce (Lactuca sativa cv. Lucy Brown) grown under greenhouse conditions and drip irrigation. Spermidine was the prevalent amine, followed by putrescine, cadaverine and agmatine. The contents of every amine, except agmatine, increased with water stress but not capable of negatively affecting the sensory quality of the lettuce. Changes in plant polyamine metabolism occur in response to a variety of abiotic stresses, however, the physiological significance of increased polyamine levels in abiotic stress responses is still unclear (Gill & Tuteja, 2010).

6.4 Cabbage

Supplying irrigation to achieve maximum cabbage (Brassica oleracea L. Capitata group) yield will also optimize sensory quality by minimizing the compounds responsible for pungency. However, glucosinolate concentration will be reduced (Radovich et al., 2005). Glucosinolates Supplying irrigation to achieve maximum cabbage (Brassica oleracea L. Capitata group) yield will also optimize sensory quality by minimizing the compounds responsible for pungency. However, glucosinolate concentration will be reduced (Radovich et al., 2005). Glucosinolates are amino acid-derived secondary metabolites that may exhibit antibiotic, anti-carcinogenic and organoleptic activity after hydrolysis. Sinigrin and progoitrin are the most important compounds with regard to flavour, since they are the primary determinants of pungency, bitterness and sulphurous aroma in cabbage (Buttery et al., 1976; Fahey et al., 2001; Talalay & Fahey, 2001).

6.5 Mustard

Similar to cabbage, mustard is a glucosinolate containing plant. This class of secondary metabolites are found almost exclusively in plants of the order Brassicales, including horticulturally important crop plants of the Brassicaceae family (Fahey et al., 2001). In order to evaluate the glucosinolate metabolism response to environmental stresses, Schreiner et al. (2009) submitted the Ethiopian mustard (Brassica carinata), lines Holeta-1 and 37-A, to water restriction treatments. Control plants were maintained at 80% of field capacity during the total growing period. Water restriction consisted of systematic decreases in soil water content (40, 23, 17, and 15% of field capacity in the 6-8, 11-12, 13-14, and 15-16 leaf stages, respectively). The most abundant glucosinolate found in both lines was 2-propenyl glucosinolate, followed by 3- indolyl methyl glucosinolate. The concentration of these two compounds remained constant in control plants from 6-8 to 15-16 leaf stage. In contrast, the stressing condition led plants to a distinct increase (80-120%) of 2-propenyl glucosinolate and 3-indolyl methyl glucosinolate in the leaves of both lines. The increases in leaf glucosinolates was inversely correlated to soil water content, in both lines, leading to severe yield losses.

6.6 Broccoli

We have demonstrated that low soil water content (0.40 MPa of soil water tension) during broccoli growth leads to leaf size reduction, without affecting weight or yield, and contributes to the maintenance of green colour, possibly due to induced cytokinin synthesis (Zaicovski et al., 2008). Cytokinins, which are considered senescence inhibitors acting as ethylene antagonists and protectors of membranes, mitochondria, and plastid metabolism are also known to be induced in response to stresses (Xu & Huang, 2009). However, severe stress (−0.6 MPa of soil water pressure) leads to negative effects on broccoli yield and morphology as observed by Wurr et al. (2002). In another study we cultivated broccoli under low (0.40 MPa) and normal (0.04 MPa, equivalent to field capacity) soil water content, stored it under low (1 °C) and room (23 °C) temperature, and assessed for changes in colour, bioactive compounds, and antioxidant activity. We concluded that low soil water content during plant growth and postharvest cold storage are the conditions that, combined, give the best preservation of colour, antioxidant activity, and l-ascorbic acid and 5-methyltetrahydrofolate contents (Cogo et al., 2011).

6.7 Eggplant

Kirnak et al. (2002) investigated the effects of deficit irrigation on fruit yield and quality in eggplant (Solanum melongena L.) cv. Pala. The experiment consisted of the following treatments: (1) well-watered, receiving 100% of plant evaporation on a daily basis (C); (2) water-stressed, receiving 90% of plant evaporation at 4-day intervals (WS1); (3) waterstressed, receiving 80% of plant evaporation at 8-day intervals (WS2); and (4) water-stressed, receiving 70% of plant evaporation at 12-day intervals (WS3). The highest yield, the largest and the heaviest fruit was observed in well-watered treatment (C). WS1 did not significantly affect fruit yield or fruit size but produced fruit slightly lighter, whilst nutrient and chlorophyll concentrations in leaves were the same as in C. WS1 presented higher soluble dry matter (SDM) than C. The WS2 and WS3 treatments caused reductions in most parameters, except SDM concentrations in the fruit, compared to C treatment. WS2 and WS3 reduced marketable yield by 12% and 28.6%, respectively, compared with C. The highest total water use efficiency and irrigation water use efficiency were in WS2, resulting in a 20.4% water saving compared with C.

7. ATTRIBUTES COMMONLY EVALUATED IN FRUIT AND VEGETABLE UNDER WATER STRESS

It became noticeable that in water stress related studies, vegetables are evaluated mainly in terms of health promoting compounds while fruit are evaluated mainly in terms of sensorial related attributes (Table 1)

Table 1. Effect of limited water supply during the plant production cycle on postharvest quality of fruit and vegetables. The most relevant reference is cited.

Plant specie	Cultivar/Var.	Healthy related metabolites	Quality related properties	Yield	Reference
Kiwifruit	Hayward	-	TSS[a] (↑) Firmness (=)	(↓)	Miller et al. (1998)
	Hayward	-	Firmness (↑)	(=)	Reid et al. (1996)
Strawberry	Elsanta	-	Sweetness (↑)	(=)	Bordonaba & Terry (2010)
Apricot	Búlida	-	TSS (↑) TA[b] (↑) Colour (↑) PHDQ[c] (↑)	(=)	Perez-Pastor (2007)
Peach	-		TSS (↑) Colour (↑) TA (↑)	(=)	Gelly et al. (2003) Gelly et al. (2004)
Pear	Conference		TSS (↑) Firmness (↑) PHDQ[c] (↑)	(↓)	Lopez et al. (2011)
Plum	Black-Gold		TSS (↑	(=)	
Orange	Salustiano	-	TSS (↑) TA (↑)	(=)	García-Tejero (2010)
	Clementina de Nules	-	TSS (↑) TA (↑)	(=)	Velez et al. (2007)
Table grape	Rizamat	Ascorbic. acid (↑)	TSS (↑) TA (↓)	(=)	Du et al. (2008)
Grapevine	Moscatel	Anthocianins (↑) Phenols (↑)	Colour (↑)	(=)	Santos et al. (2007)
Tomato	Zarina	Flavonoids (↑)	-	-	Sánchez-Rodríguez et al. (2011)
Cabbage	Holeta-1 and 37-A	Glucosinolates (↑)	-	(↓)	Schreiner et al. (2009)
Broccoli	Green star	Ascorbic. Acid (↑) Folates (↑)	PHDQ (↑)	(=)	(Cogo et al., 2011)

[a] TSS: Total soluble solids.
[b] TA: Titratable Acidity.
[c] PHDQ: Post-harvest desirable qualities.

8. CONCLUSION

Environmental sustainability is a current issue in global agenda. This concept comprises, among other concerns, the efficient use of water. Climate change is a reality and the pressure over water reserves will increase in next years, as well over high water consuming activities. Agricultural practices are among the biggest water consuming activities, considering that, alternatives to reduce water use in agricultural practices is of special interest in the present moment. In this review we exposed successful experiences with the intent to reduce water use with minimal losses in yield and quality. Irrigation practices often aims at total replacement of culture evapotranspiration in order to obtain maximum yield. In many occasions experiments demonstrate that is possible to reduce water use without significant losses in yield. In addition, the increases in health/quality related compounds and postharvest preservation are evident in response to environmental stress. The use of regulated stress (water stress, salinity, heat, cold, UV radiation) is a feasibly strategy to enhance accumulation of health promoting compounds in food. Another interesting perspective is the improvement of plant resistance against biotic stresses (e.g. pests and diseases) when submitted to controlled abiotic stresses, as was scientifically demonstrated in experiments presented in this review. However, more studies must be performed to determine the effects of different water stresses on different edible plants.

9. ACKNOWLEDGMENT

The authors acknowledge CAPES and CNPq from Brazilian government for the financial support.

10. REFERENCES

1. Bejarano, L., Mignolet, E., Devaux, A., Espinola, N., Carrasco, E., Larondelle, Y. (2000). Glycoalkaloids in potato tubers: the effect of variety and drought stress on the α- solanine and α-chaconine contents of potatoes. Journal of the Science of Food and Agriculture, Vol. 80, No. 14, pp. 2096-2100, ISSN 1097-0010

2. Bordonaba, J.G. & Terry, L.A. (2010). Manipulating the taste-related composition of strawberry fruits (Fragaria x ananassa) from different cultivars using deficit irrigation. Food Chemistry, Vol. 122, No. 4, (Oct), pp. 1020-1026, ISSN 0308-8146

3. Buendía, B., Allende, A., Nicolás, E., Alarcón, J.J., Gil, M.I. (2008). Effect of Regulated Deficit Irrigation and Crop Load on the Antioxidant Compounds of Peaches. Journal of Agricultural and Food Chemistry, Vol. 56, No. 10, pp. 3601-3608, ISSN 0021-8561

4. Buttery, R.G., Guadagni, D.G., Ling, L.C., Seifert, R.M., Lipton, W. (1976). Additional volatile components of cabbage, broccoli, and cauliflower. Journal of Agricultural and Food Chemistry, Vol. 24, No. 4, (1976), pp. 829-832, ISSN 0021-8561

5. Chu, H.Y., Wegel, E., Osbourn, A. (2011). From hormones to secondary metabolism: the emergence of metabolic gene clusters in plants. Plant Journal, Vol. 66, No. 1, (Apr), pp. 66-79, ISSN 0960-7412

6. Coelho, A.F.S., Gomes, E.P., Sousa, A.D., Gloria, M.B. (2005). Effect of irrigation level on yield and bioactive amine content of American lettuce. Journal of the Science of Food and Agriculture, Vol. 85, No. 6, (Apr), pp. 1026-1032, ISSN 0022-5142

7. Cogo, S.L.P., Chaves, F.C., Schirmer, M.A., Zambiazi, R.C., Nora, L., Silva, J.A., Rombaldi, C.V. (2011). Low soil water content during growth contributes to preservation of green colour and bioactive compounds of cold-stored broccoli (Brassica oleraceae L.) florets. Postharvest Biology and Technology, Vol. 60, No. 2, pp. 158-163, ISSN 0925-5214

8. Costa, J.M., Ortuno, M.F., Chaves, M.M. (2007). Deficit irrigation as a strategy to save water: Physiology and potential application to horticulture. Journal of Integrative Plant Biology, Vol. 49, No. 10, (Oct), pp. 1421-1434, ISSN 1672-9072

9. D'Auria, J.C. & Gershenzon, J. (2005). The secondary metabolism of Arabidopsis thaliana: growing like a weed. Current Opinion In Plant Biology, Vol. 8, No. 3, (Jun), pp. 308-316

10. Dixon, R.A. & Paiva, N.L. (1995). Stress-Induced Phenylpropanoid Metabolism. Plant Cell, Vol. 7, No. 7, (Jul), pp. 1085-1097

11. Du, T., Kang, S., Zhang, J., Li, F., Yan, B. (2008). Water use efficiency and fruit quality of table grape under alternate partial root-zone drip irrigation. Agricultural Water Management, Vol. 95, No. 6, pp. 659-668, ISSN 0378-3774

12. Fahey, J.W., Zalcmann, A.T., Talalay, P. (2001). The chemical diversity and distribution of glucosinolates and isothiocyanates among plants. Phytochemistry, Vol. 56, No. 1, (Jan), pp. 5-51, ISSN 0031-9422

13. Friedman, M. (2006). Potato glycoalkaloids and metabolites: Roles in the plant and in the diet. Journal of Agricultural and Food Chemistry, Vol. 54, No. 23, (Nov 15), pp. 8655-8681

14. García-Tejero, I., Jimenez-Bocanegra, J.A., Martinez, G., Romero, R., Duran-Zuazo, V.H., Muriel-Fernandez, J. (2010). Positive impact of regulated deficit irrigation on yield and fruit quality in a commercial citrus orchard Citrus sinensis (L.) Osbeck, cv. salustiano. Agricultural Water Management, Vol. 97, No. 5, (May), pp. 614-622, ISSN 0378-3774

15. Gill, S.S. & Tuteja, N. (2010). Polyamines and abiotic stress tolerance in plants. Plant Signaling & Behavior, Vol. 5, No. 1, pp. 26-33

16. Godfray, H.C.J., Crute, I.R., Haddad, L., Lawrence, D., Muir, J.F., Nisbett, N., Pretty, J., Robinson, S., Toulmin, C., Whiteley, R. (2010). The future of the global food system. Philosophical

Transactions of the Royal Society B: Biological Sciences, Vol. 365, No. 1554, (September 27, 2010), pp. 2769-2777

17. Hooper, L. & Cassidy, A. (2006). A review of the health care potential of bioactive compounds. Journal of the Science of Food and Agriculture, Vol. 86, No. 12, pp. 1805-1813, ISSN 1097-0010

18. Intrigliolo, D.S. & Castel, J.R. (2010). Response of plum trees to deficit irrigation under two crop levels: tree growth, yield and fruit quality. Irrigation Science, Vol. 28, No. 6, (Sep), pp. 525-534, ISSN 0342-7188

19. Kirnak, H., Tas, I., Kaya, C., Higgs, D. (2002). Effects of deficit irrigation on growth, yield, and fruit quality of eggplant under semi-arid conditions. Australian Journal of Agricultural Research, Vol. 53, No. 12, (2002), pp. 1367-1373, ISSN 0004-9409

20. Krouk, G., Ruffel, S., Gutierrez, R.A., Gojon, A., Crawford, N.M., Coruzzil, G.M., Lacombe, B. (2011). A framework integrating plant growth with hormones and nutrients. Trends In Plant Science, Vol. 16, No. 4, (Apr), pp. 178-182, ISSN 1360-1385

21. Lopez, G., Larrigaudiere, C., Girona, J., Behboudian, M.H., Marsal, J. (2011). Fruit thinning in 'Conference' pear grown under deficit irrigation: Implications for fruit quality at harvest and after cold storage. Scientia Horticulturae, Vol. 129, No. 1, (May), pp. 64-70, ISSN 0304-4238

22. Miller, S.A., Smith, G.S., Boldingh, H.L., Johansson, A. (1998). Effects of water stress on fruit quality attributes of kiwifruit. Annals of Botany, Vol. 81, No. 1, pp. 73-81, ISSN 0305- 7364

23. Mpelasoka, B.S., Behboudian, M.H., Dixon, J., Neal, S.M., Caspari, H.W. (2000). Improvement of fruit quality and storage potential of 'Braeburn' apple through deficit irrigation. Journal of Horticultural Science & Biotechnology, Vol. 75, No. 5, (Sep), pp. 615-621, ISSN 0022-1589

24. Mpelasoka, B.S., Behboudian, M.H., Ganesh, S. (2001a). Fruit quality attributes and their interrelationships of 'Braeburn' apple in response to deficit irrigation and to crop load. Gartenbauwissenschaft, Vol. 66, No. 5, (Sep-Oct), pp. 247-253, ISSN 0016-478X

25. Mpelasoka, B.S., Behboudian, M.H., Mills, T.M. (2001b). Effects of deficit irrigation on fruit maturity and quality of 'Braeburn' apple. Scientia Horticulturae, Vol. 90, No. 3-4, (Nov), pp. 279-290, ISSN 0304-4238

26. Mpelasoka, B.S., Behboudian, M.H., Mills, T.M. (2001c). Water relations, photosynthesis, growth, yield and fruit size of 'Braeburn' apple: Responses to deficit irrigation and to crop load. Journal of Horticultural Science & Biotechnology, Vol. 76, No. 2, (Mar), pp. 150-156, ISSN 0022-1589

27. Osbourn, A. (2010). Secondary metabolic gene clusters: evolutionary toolkits for chemical innovation. Trends in Genetics, Vol. 26, No. 10, (Oct), pp. 449-457, ISSN 0168-9525 Peleg, Z. & Blumwald, E. (2011). Hormone balance and abiotic stress tolerance in crop plants. Current Opinion In Plant Biology, Vol. 14, No. 3, pp. 290-295, ISSN 1369-5266

28. Perez-Pastor, A., Ruiz-Sanchez, M.C., Martinez, J.A., Nortes, P.A., Artes, F., Domingo, R. (2007). Effect of deficit irrigation on apricot fruit quality at harvest and during storage. Journal of the Science of Food and Agriculture, Vol. 87, No. 13, (Oct), pp. 2409-2415, ISSN 0022-5142

29. Perez-Sarmiento, F., Alcobendas, R., Mounzer, O., Alarcon, J., Nicolas, E. (2010). Effects of regulated deficit irrigation on physiology and fruit quality in apricot trees. Spanish Journal of Agricultural Research, Vol. 8, (2010), pp. S86-S94, ISSN 1695-971X

30. Pospíšilová, J. & Dodd, I.C. (2005). Role of plant growth regulators in stomatal limitation to photosynthesis during water stress, In Handbook of Photosynthesis, M. Pessarakli, (Ed.), 811-825, CRC Press, ISBN 0824758390, New York, USA

31. Proietti, S., Rouphael, Y., Colla, G., Cardarelli, M., De Agazio, M., Zacchini, M., Rea, E., Moscatellol, S., Battistelli, A. (2008). Fruit quality of mini-watermelon as affected by grafting and irrigation regimes. Journal of the Science of Food and Agriculture, Vol. 88, No. 6, (Apr 30), pp. 1107-1114, ISSN 0022-5142

32. Radovich, T.J.K., Kleinhenz, M.D., Streeter, J.G. (2005). Irrigation timing relative to head development influences yield components, sugar levels, and glucosinolate concentrations in cabbage. Journal of the American Society for Horticultural Science, Vol. 130, No. 6, (Nov), pp. 943-949, ISSN 0003-1062

33. Reid, J.B., Brash, D.W., Sorensen, I.B., Bycroft, B. (1996). Improvement in kiwifruit storage life caused by withholding early-season irrigation. New Zealand Journal of Crop and Horticultural Science, Vol. 24, No. 1, (Mar), pp. 21-28, ISSN 0114-0671

34. Sánchez-Rodríguez, E., Moreno, D.A., Ferreres, F., Rubio-Wilhelmi, M.d.M., Ruiz, J.M. (2011). Differential responses of five cherry tomato varieties to water stress: Changes on phenolic metabolites and related enzymes. Phytochemistry, Vol. 72, No. 8, pp. 723-729, ISSN 0031-9422

35. Santner, A. & Estelle, M. (2009). Recent advances and emerging trends in plant hormone signalling. Nature, Vol. 459, No. 7250, (June 2009), pp. 1071-1078, ISSN 0028-0836

36. Santos, T.P.d., Lopes, C.M., Rodrigues, M.L., Souza, C.R.d., Ricardo-da-Silva, J.M., Maroco, J.P., Pereira, J.S., Chaves, M.M. (2007). Effects of deficit irrigation strategies on cluster microclimate for improving fruit composition of Moscatel field-grown grapevines. Scientia Horticulturae, Vol. 112, No. 3, pp. 321-330, ISSN 0304-4238

37. Schreiner, M., Beyene, B., Krumbein, A., Stuetzel, H. (2009). Ontogenetic Changes of 2- Propenyl and 3-Indolylmethyl Glucosinolates in Brassica carinata Leaves as Affected by Water

Supply. Journal of Agricultural and Food Chemistry, Vol. 57, No. 16, (Aug 26), pp. 7259-7263, ISSN 0021-8561

38. Stefanelli, D., Goodwin, I., Jones, R. (2010). Minimal nitrogen and water use in horticulture: Effects on quality and content of selected nutrients. Food Research International, Vol. 43, No. 7, (Aug), pp. 1833-1843, ISSN 0963-9969

39. Stoll, M., Loveys, B., Dry, P. (2000). Hormonal changes induced by partial rootzone drying of irrigated grapevine. Journal Of Experimental Botany, Vol. 51, No. 350, (Sep), pp. 1627- 1634, ISSN 0022-0957

40. Talalay, P. & Fahey, J.W. (2001). Phytochemicals from cruciferous plants protect against cancer by modulating carcinogen metabolism. Journal Of Nutrition, Vol. 131, No. 11, (Nov), pp. 3027S-3033S, ISSN 0022-3166

41. Terry, L.A., Chope, G.A., Bordonaba, J.G. (2007). Effect of water deficit irrigation and inoculation with Botrytis cinerea on strawberry (Fragaria x ananassa) fruit quality. Journal of Agricultural and Food Chemistry, Vol. 55, No. 26, (Dec 26), pp. 10812-10819, ISSN 0021-8561

42. Thipyapong, P., Melkonian, J., Wolfe, D.W., Steffens, J.C. (2004). Suppression of polyphenol oxidases increases stress tolerance in tomato. Plant Science, Vol. 167, No. 4, (Oct), pp. 693-703, ISSN 0168-9452

43. Treeby, M.T., Henriod, R.E., Bevington, K.B., Milne, D.J., Storey, R. (2007). Irrigation management and rootstock effects on navel orange Citrus sinensis (L.) Osbeck fruit quality. Agricultural Water Management, Vol. 91, No. 1-3, (Jul 16), pp. 24-32, ISSN 0378- 3774

44. Velez, J.E., Intrigliolo, D.S., Castel, J.R. (2007). Scheduling deficit irrigation of citrus trees with maximum daily trunk shrinkage. Agricultural Water Management, Vol. 90, No. 3, (Jun), pp. 197-204, ISSN 0378-3774

45. Wang, Y. & Frei, M. (2011). Stressed food - The impact of abiotic environmental stresses on crop quality. Agriculture, Ecosystems & Environment, Vol. 141, No. 3-4, pp. 271-286, ISSN 0167-8809

46. Wilkinson, S. & Davies, W.J. (2010). Drought, ozone, ABA and ethylene: new insights from cell to plant to community. Plant Cell and Environment, Vol. 33, No. 4, (April 2010), pp. 510-525, ISSN 0140-7791

47. Wink, M. (1999). Introduction: biochemistry, role and biotechnology of secondary metabolites, In Biochemistry, role and biotechnology of secondary metabolites, M. Wink, (Ed.), 1-16, Sheffield Academic Press and CRC Press, Annual Plant Reviews, ISBN 978-1-4051- 8528-8, Sheffield, UK

48. Wurr, D.C.E., Hambidge, A.J., Fellows, J.R., Lynn, J.R., Pink, D.A.C. (2002). The influence of water stress during crop growth on the postharvest quality of broccoli. Postharvest Biology and Technology, Vol. 25, No. 2, (Jun), pp. 193-198

49. Xu, Y. & Huang, B.R. (2009). Effects of foliar-applied ethylene inhibitor and synthetic cytokinin on creeping bentgrass to enhance heat tolerance. Crop Science, Vol. 49, No. 5, (SepOct), pp. 1876-1884, ISSN 0011-183X

50. Yamaguchi-Shinozaki, K. & Shinozaki, K. (2006). Transcriptional regulatory networks in cellular responses and tolerance to dehydration and cold stresses, In Annual Review Of Plant Biology, D.P. Delmer, (Ed.), 781-803, Annual Reviews, ISBN 1543-5008, New York, USA

51. Zaicovski, C.B., Zimmerman, T., Nora, L., Nora, F.R., Silva, J.A., Rombaldi, C.V. (2008). Water stress increases cytokinin biosynthesis and delays postharvest yellowing of broccoli florets. Postharvest Biology and Technology, Vol. 49, No. 3, (Sep), pp. 436-439, ISSN 0925- 5214

12

SOUND WAVES DELAY TOMATO FRUIT RIPENING BY NEGATIVELY REGULATING ETHYLENE BIOSYNTHESIS AND SIGNALING GENES

Joo-Yeol Kim[a], Jin-Su Lee[b], Taek-Ryoun Kwon[a], Soo-In Lee[a], Jin-A. Kim[a], Gyu-Myoung Lee[a], Soo-Chul Park[a], Mi-Jeong Jeong[a]

[a]Department of Agricultural Biotechnology, National Academy of Agricultural Science (NAAS), 370 Nongsaengmyoeng-ro, Wansan-gu, Jeonju, Jeollabuk-do 560-500, South Korea

[b] Postharvest Research Team, National Institute of Horticultural and Herbal Science (NIHHS), 100, Nongsaengmyeong-ro, Iseo-myeon, Wanju-gun, Jeollabuk-do 565-852, South Korea

ABSTRACT

Regulation of tomato fruit ripening may help extend fruit shelf life and prevent losses due to spoilage. Here, tomato fruit were investigated whether sound treatment could delay their ripening. Harvested fruit were treated with low-frequency sound waves (1 kHz) for 6 h, and then monitored various characteristics of the fruit over 14-days at 23 ± 1 °C. Seven days after the treatment, 85% of the treated fruit were green, versus fewer than 50% of the non-treated fruit. Most of the tomato fruit had transitioned to the red ripening stage by 14 days after treatment. Ethylene production and respiration rate were lower in the sound-treated than non-treated tomatoes. Furthermore, changes in surface color and flesh firmness were delayed in the treated fruit. To investigate how sound wave treatment effects on fruit ripening, the expression of ethylene-related genes was analyzed by quantitative real-time RT-PCR analysis. The expression level of several ethylene biosynthetic (*ACS2, ACS4, ACO1, E4* and *E8*) and ripening-regulated (*RIN, TAGL1, HB-1, NOR, CNR*) genes was influenced by sound wave treatment. These results indicated that sound wave treatment delays tomato fruit ripening by altering the expression of important genes in the ethylene biosynthesis and ethylene signaling pathways.

Keywords: Sound waves; Tomato; Ethylene biosynthesis; Fruit ripening; Ethylene signaling

1. INTRODUCTION

Plants are frequently exposed to abiotic and biotic environmental stresses. Stress signals are transmitted to cells, where they activate various stress response mechanisms. The mechanism by which plants recognize external signals and then alter the expression of genes to affect physiological and/or metabolic processes is complex. Much research has examined the signal transduction mechanisms that regulate the defense or growth responses of plants following artificial stimulation, such as mechanical stimulation (Lin et al., 2009, Braam, 2004 and Monshausen and Gilroy, 2009). Another such external signal is sound.

Previous reports on the effects of sound on plants failed to specify the exact experimental conditions (Weinberger and Measures, 1978). Sound vibrations stimulate seed germination and plant growth and increase the rate of hypocotyl elongation in *Oryza sativa* (rice) and *Cucumis sativus* (cucumber) seedlings (Takahashi et al., 1992) and in Arabidopsis (Johnson et al., 1998). Jomdecha and Prateepasen (2011) showed that low-intensity ultrasonic sound enhances the growth of *Saccharomyces cerevisiae.*Furthermore, transgenic rice plants harboring an *Alcohol dehydrogenase* (*Ald*) promoter:*GUS* reporter construct showed a significant increase in *GUS* expression after sound treatment at 125 Hz, 250 Hz, or 1 kHz. By contrast, these plants showed a significant decrease in *GUS* mRNA levels after a treatment of 50 Hz, suggesting that the*Ald* promoter responds to sound in a frequency-specific manner (Jeong et al., 2008). In addition, through biochemical analyses, Bochu et al. (2004) reported that sound wave stimuli (1.4 kHz, 95 dB, 10 days) were associated with increased levels of indole acetic acid (IAA) and decreased levels of abscisic acid (ABA) in Chrysanthemum. Accordingly, sound waves are thought to regulate a variety of plant hormones and genes.

Fruits are divided into climacteric and non-climacteric types based on their respiration patterns during ripening. Climacteric fruits, such as tomato, apple, avocado, melon, pear, kiwifruit, and peach, exhibit increased respiration and ethylene production during ripening. However, respiration and ethylene production rates barely change during ripening of non-climacteric fruit, such as grapes, cherry, strawberry, pineapple, and citrus (Lin et al., 2009 and Cara and Giovannoni, 2008). Almost all fruits start aging rapidly when ethylene production increases and once fruit have ripened their quality declines. Suppression of senescence in fruit results in a longer shelf life. Tomato is a major model system for the study of the molecular mechanisms underlying fruit development and the role of ethylene in climacteric fruit ripening. It has various useful characteristics, such as efficient genetic transformation, distinct ripening phenotypes, small genome size, short generation time, and simple diploid genetics (Liu et al., 2014 and Giovannoni, 2004).

Ethylene is a plant hormone involved in fruit, flower, and leaf senescence. The ethylene biosynthesis precursor S-adenosylmethionine (SAM) is converted into 1-aminocyclopropane-1-carboxylic acid (ACC) by ACC synthase (ACS), and ACC oxidase (ACO) then catalyzes the synthesis of ethylene (Yang and Hoffman, 1984 and Johnson and Ecker, 1998). Ethylene biosynthesis in tomato fruit can be divided into systems 1 and 2 during climacteric fruit ripening (Lelièvre et al., 1997 and Alexander and Grierson, 2002). System 1 is auto-inhibitory and functions during normal growth and development and the stress response, whereas system 2 operates during floral and leaf senescence and fruit ripening (Katz et al., 2004 and Barry and Giovannoni, 2007).

Expression of *LeACS2*, *LeACS4,* and *LeACO1* increases significantly during fruit ripening, indicating that these genes are major regulators of ripening (Hoogstrate et al., 2014 and Lincoln et al., 1993). The ethylene-inducible genes *E4* and *E8* also have important functions during fruit ripening. Transcription of *E4* is rapidly activated by ethylene in both leaves and fruit. By contrast, expression of *E8* is highly induced by ethylene in tomato fruit, but is not induced in response to ethylene in leaves. Although *E4*and *E8* are coordinately expressed during fruit ripening, *E8* is controlled both by ethylene and by separate developmental signals, whereas *E4* expression is strictly dependent on ethylene (Kesanakurti et al., 2012 and Lincoln and Fischer, 1988). Furthermore,*TOMATO AGAMOUS-LIKE 1* (*TAGL1*) MADS box gene as a transcription factor plays an important function in both fruit expansion and ripening. Silencing of *TAGL1* does not affect floral organs, but affects ripening of the tomato fruit (Vrebalov et al., 2009,Giménez et al., 2010 and Pan et al., 2010). *RIPENING-INHIBITOR (RIN)* functions upstream of ethylene in the ripening cascade and determine the competence of fruit to ripen. *RIN* was shown to regulate *LeACS2* and *LeACS4* expression (Vrebalov et al., 2002). Consequently, *TAGL1* and *RIN* affect ethylene metabolism (Yokotani et al., 2004). In addition, virus-induced gene silencing of *LeHB-1*, which encodes an HD-Zip homeobox protein, reduced *LeACO1* mRNA expression and inhibited ripening in tomato fruit (Lin et al., 2008). In this study, tomato fruits were investigated whether sound wave treatment could delay ripening by regulating the expression of genes involved in ethylene biosynthesis and signaling.

2. MATERIALS AND METHODS

2.1. Plant materials and treatment conditions

Tomato (*Solanum lycopersicum* L. cv. Dotaerang) fruits were sampled at the mature green stage of development (average fruit weight 172.9 ± 1.7 g) from a commercial greenhouse facility at Yong-in, Korea. After transport to the laboratory of the Rural Development

Administration on the day of harvest, the fruits were exposed to a 1 kHz sound wave for 6 h. The single-frequency signal was generated using Pro Tools M-Powered software (Avid Technology, USA). The speaker volume was 100 dB. To prevent extraneous noise, experiments were conducted inside a custom-made, sound-proof chamber (Korea Scientific Technique Industry Co., Korea). The sound level within the chamber was approximately 40 dB, while the sound level in a commercial growth chamber reaches generally about 80 dB. Tomato fruits were placed in the sound-proof chambers to prevent transfer of vibrations between samples during the sound wave treatments. The sound wave treatment was conducted at 23 ± 1 °C. Control groups were achieved with the exposure of fruit in the same conditions except for the sound interference. Fruit samples were placed in a storage room at 23 ± 1 °C and monitored for quality changes after sound wave treatments.

2.2. Ethylene production and respiration rate determination

Thirty-two tomato fruits per treatment were used for assessing ethylene production and respiration rate at each measurement interval. Eight tomato fruits were weighed and sealed in a 3900 mL plastic container ($n = 4$) equipped with a septum for 1 h. Headspace samples (1 mL) were withdrawn from the container and ethylene and carbon dioxide concentrations were analyzed using a gas chromatograph (Varian 450, Varian Inc., CA, USA) equipped with a flame ionization detector (FID) and thermal conductivity detector (TCD).

Ethylene levels were measured by a FID equipped with an active alumina 60–80 mesh column. The injector, oven, and detector were maintained at 110, 70, and 250 °C, respectively. Respiration rate was determined based on carbon dioxide levels using a TCD equipped with an active carbon 60–80 mesh column. The injector, oven, and detector were maintained at 110, 70, and 150 °C, respectively. Helium (30 mL min^{-1}) was used as the carrier gas for ethylene and carbon dioxide detection.

2.3. Fruit quality parameters

Surface color was measured using a Chromameter (CR-400, Konica Minolta Sensing Inc., Osaka, Japan). Two equatorial regions of each fruit were marked and the surface color of 30 fruit per treatment ($n = 30$) was assessed repeatedly at each measurement interval. The hunter '*a*' value ranging from green to red was used to determine surface color change during storage.

Flesh firmness was determined using a texture analyzer (TA plus, Ametek Lloyd Instruments Ltd., Hampshire, UK) equipped with a 5 mm cylindrical probe and 250 N load cell. The equatorial region of each fruit was cut into 20-mm thick slices and flesh firmness was evaluated. The probe was pressed 10 mm into the mesocarp surface of a slice with a crosshead speed of 2 mm s^{-1} and the positive peak force (*N*) was recorded. Flesh firmness of slices derived from 20 fruits per treatment ($n = 20$) were evaluated at each measurement interval.

2.4. RNA extraction and quantitative real-time PCR (qRT-PCR)

Tomato fruits were obtained from at least three individuals per step. Directly after harvesting, the fruits were frozen in liquid nitrogen and stored at −80 °C. These fruits were ground into powder with pestle and mortar in liquid nitrogen to keep the tissue frozen. Total RNA was extracted using a Plant RNeasy Extraction Kit (Qiagen). Total RNA was treated with DNase I (Qiagen) and cDNA was generated using ampfiRivert Platinum cDNA Synthesis Master Mix (GenDEPOT, South Korea). Quantitative real-time RT-PCR (qRT-PCR) was performed using SYBR Premix Ex Taq (Takara) and the StepOnePlus Real-Time PCR System (Applied Biosystems). The relative mRNA levels were determined by normalizing the PCR threshold cycle number of each gene with that of the *Ubiquitin3* reference gene. Three biological replicates were used in the experiments. The primers used for PCR are shown in Table S1.

3. RESULTS AND DISCUSSION

3.1. Sound wave treatment delayed ripening in tomato fruit

Sound wave significantly effects on the ripening process of tomato fruit. To evaluate the impact of sound waves on fruit ripening, tomato fruits at the mature green stage were treated with sound waves of 0 Hz, 250 Hz, 500 Hz, 800 Hz, 1 kHz, and 1.5 kHz for 6 h. For all of the sound wave treatments except for 800 Hz and 1.5 kHz, ripening was delayed as compared with the non-treated control (Fig. 1). Since the 1 kHz sound wave showed the strongest effect on ripening process, tomato fruit were treated with 1 kHz sound for phenotypic analysis over 14 days of storage at 23 ± 1 °C. Thus, most of the results in this study were obtained using tomato fruit treated with 1 kHz, unless otherwise indicated. Under these conditions, the mature green stage tomato fruit transitioned to the breaker stage after 2 days. Tomato fruit treated with sound waves of 1 kHz exhibited delayed ripening as compared with the non-treated fruit at 5 or 7 days after treatment (Fig. 2A). Seven days after treatment, over 85% of the treated fruit was still green, whereas less than 50% of the non-treated tomato fruit were green (Fig. 2B). Finally, most of the tomato fruit transitioned to the red ripening stage by 14 days after treatment. These results indicate that sound wave treatment delays tomato fruit ripening.

These findings could be explained partially with the previous reports on the effects of sound wave to the deformability of plant cell. Sound wave exposure changes the deformability of the plant cell membrane, and the deformability is closely related to the frequency of the sound wave (Bochu et al., 2001).

3.2. Sound wave treatment reduced ethylene production and respiration rate in tomato fruit

External stimuli like sound wave also could regulate various biological process and genes. A scientific report by Johnson et al. (1998) reveals that mechanotransduction occurs in cells and tissues, and external mechanical stresses stimulate various signal transduction mechanisms. In this study, external stimuli with the sound wave of 1 kHz applied to 32 tomato fruits at various time points after treatment (1–15 days) prior to the determination of ethylene production. As shown in Fig. 3A, ethylene production was lower in tomatoes treated with sound waves of 1 kHz than in the non-treated control. The ethylene production was observed the greatest reduction at 12 days after treatment. These results indicate that the imposition of the sound treatment wave significantly effects on the biological process of ethylene, resulting in the delay of the fruit ripening. These findings agree well with the previous postharvest physiology of tomato fruit suggested by Lin et al., (2009) and Cara and Giovannoni (2008). The ripening of tomato fruit, as a climacteric type fruit, is associated with increased respiration and ethylene production.

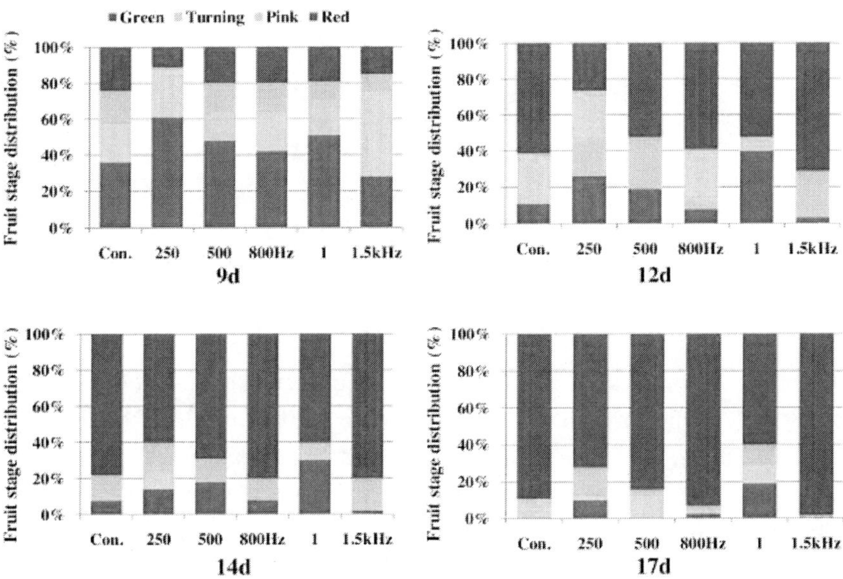

Figure 1. Comparison of fruit surface color change in response to various sound treatments in tomato. Phenotypic analysis of tomato fruit ripening at 9, 12, 14, and 17 days after a 6-h sound treatment (non-treatment, 250 Hz, 500 Hz, 800 Hz, 1 kHz, and 1.5 kHz). Fruit were maintained at 23 °C throughout the experiment. The average results of three experiments are shown.

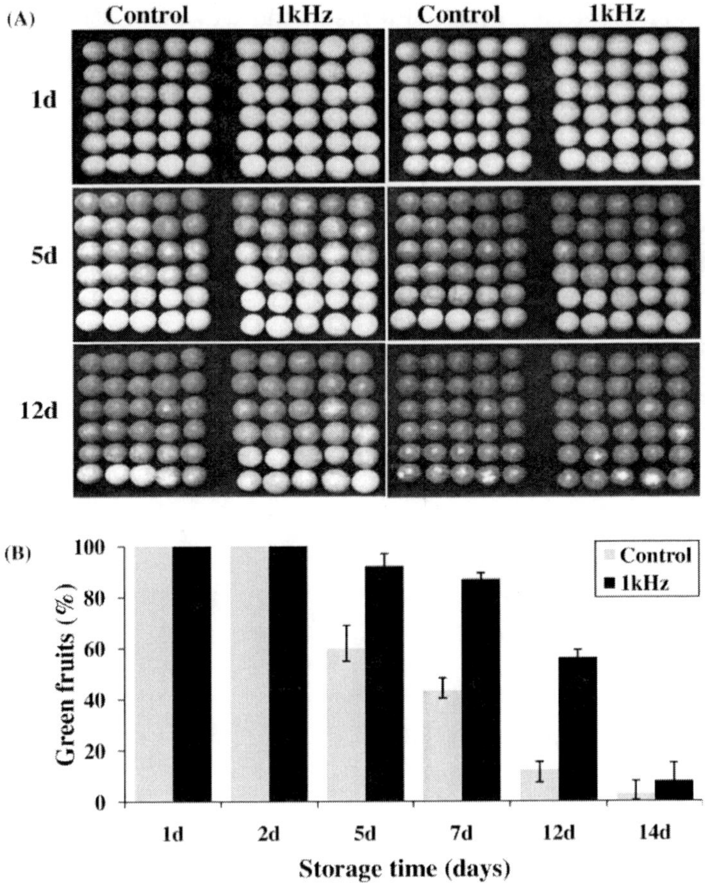

Figure 2. Ripening delay in tomato fruit subjected to a sound treatment of 1 kHz.(A) Phenotypic analysis of tomato fruit ripening at 1, 2, 5, 7, 12, and 14 days (at 23 ± 1 °C) after sound treatment (1 kHz), (B) the percentage of green tomato fruit during the 14 days (at 23 ± 1 °C) following sound treatment (1 kHz). Error bars represent SD of three biological replicates.

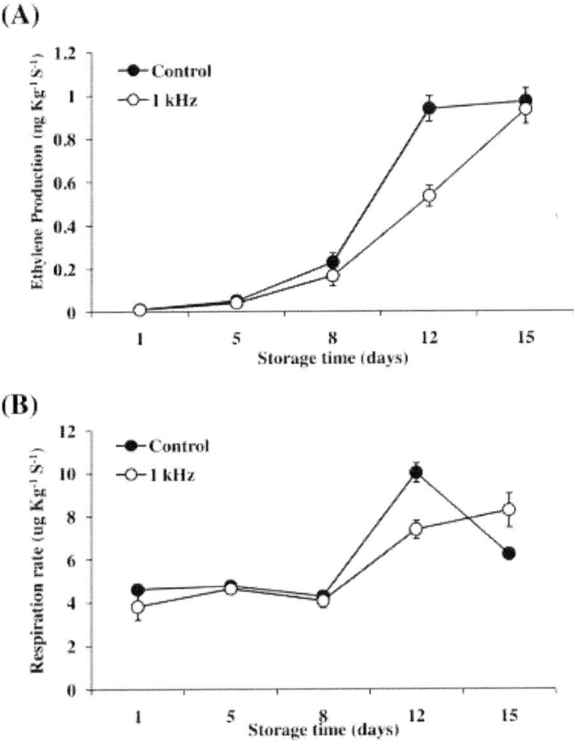

Figure 3. The effect of sound treatment on ethylene production and respiration rate in tomato fruit. (A) Changes in ethylene production of tomato fruit during storage at 23 °C after sound treatment (1 kHz). (B) Changes in respiration rate of tomato fruit during storage at 23 °C after sound wave treatment (1 kHz). Error bars represent SD of three biological replicates.

Next, the respiration rate was measured in tomato fruit treated or not with sound waves. To examine the respiration rate, the evolution rate of carbon dioxide was analyzed using a gas chromatography. The respiration rate was lower in the treated tomato fruit one day after treatment than in the non-treated control, and the greatest relative reduction was observed at 12 days after treatment (Fig. 3B). In the non-treated fruit, the respiration rate peaked at 12 days, and then decreased rapidly. At this stage, the ripening stage seemed to be completed, and senescence seemed to be in progress. By contrast, the treated tomato fruit appeared still to be ripening, because the respiration rate increased even at 15 days after treatment. These results indicate that sound wave treatment (1 kHz) delays ripening of tomato fruit by controlling the ethylene production as well as respiration rate.

3.3. Sound wave caused changes in surface color and hardness in tomato fruit

Surface color is the most important external characteristic for visual assessing of ripeness in tomato fruit. The degree of ripening is usually estimated using color charts. Surface color analysis using a chromameter provides an objective way to quantitatively measure and compare the color of tomato fruit by standardizing the elements necessary to view color. The hunter '*a*' value, which denotes colors ranging from green to red, was used to determine surface color changes during storage.

Tomato fruits were treated with 1 kHz sound waves for 6 h and then measured the surface color at various time points during the next 16 days (at 23 ± 1 °C). Changes in surface color from green to red were delayed in the treated tomato fruit compared with the non-treated fruit (Fig. 4A).

To further determine the effect of sound wave treatment on tomato fruit ripening, examined whether sound wave treatment influenced flesh firmness. The flesh firmness of non-treated tomato fruit decreased sharply after 5 days-long storage in this study, whereas that of fruit treated with sound waves decreased at a more gradual rate (Fig. 4B). These results demonstrate that sound wave treatment helped to maintain their flesh firmness.

Our study showed that the ethylene production and respiration rate were significantly reduced in plants exposed to a 1 kHz sound wave treatment (Fig. 3A and B). In addition, changes in surface color from green to red were delayed in sound wave-treated tomato fruit (Fig. 4A) and the flesh was firmer in the treated tomato fruit than in the non-treated fruit (Fig. 4B).

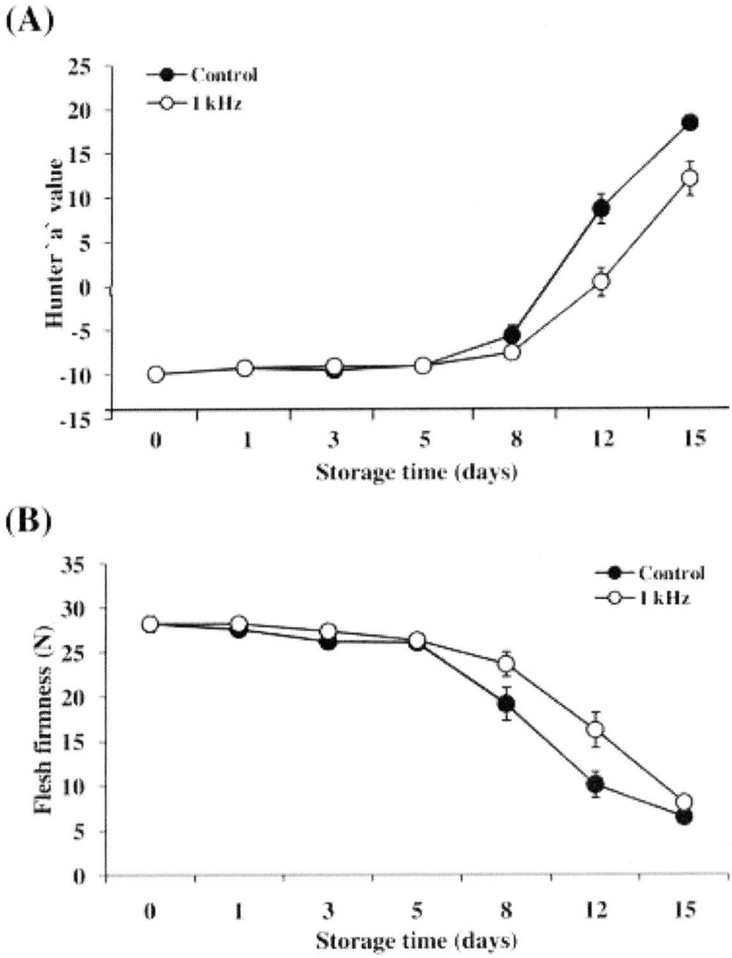

Figure 4. Sound treatment causes changes in surface color and hardness in tomato fruit. (A) Changes in Hunter '*a*' values of tomato fruit surfaces during storage at 23 °C after sound signal treatment. a* (red–green) axis: positive values are red; negative values are green; 0 is neutral. (B) Changes in flesh firmness of tomato fruit during storage at 23 °C after sound signal treatment. Samples were measured in biological triplicate. Error bars represent SD.

Thus, sound wave treatment delayed the ripening of tomato fruit (Fig. 1 and Fig. 2). Because the color change was delayed and ethylene production and rate of respiration were reduced in the sound wave-treated tomato fruit, the expression of genes involved in ethylene biosynthesis was analyzed.

3.4. Sound wave treatment affects ethylene biosynthesis-related genes

This study found that sound wave controls the expression of genes involved in the biochemical processes associated with ripening. Ethylene mediates certain developmental processes, most notably early fruit development and ripening. Having established that sound wave treatment delayed tomato fruit ripening (Fig. 1, Fig. 2, Fig. 3 and Fig. 4), tomato fruit ripening assumed to be influenced ethylene-related responses by sound wave treatment. Therefore, the expression pattern of major genes involved in ethylene biosynthesis was analyzed after sound wave treatment. The transcript levels of ethylene biosynthetic genes in sound wave-treated tomato fruit was examined at various time points after treatment (at 23 ± 1 °C) via quantitative real-time RT-PCR analysis with the gene-specific primers listed in Table S1. Under normal conditions, most of the ethylene biosynthetic genes and ripening regulated genes in tomato were expressed at low levels in green fruit, but the transcription levels of each of these genes increased at the onset of ripening, as the fruit started to produce ethylene (Fig. 5). As shown in Fig. 5, in compare to control, the expressions of *LeACS2, LeACS4* and *LeACO1* (ethylene biosynthesis-related genes) were reduced in sound wave treated tomatoes. The difference of expression level of these genes between treated and untreated groups substantially increased from 7 to 14 days. The expression level of *E4* and *E8* genes (ethylene-inducible genes) were also significantly reduced by the sound wave treatment. These genes associated with ethylene biosynthesis exhibited the most striking changes in expression between 5 and 7 days after the sound treatment. In particular, the expression of *LeACS2, LeACS4,* and *LeACO1* was markedly reduced by sound treatment, and this reduction in expression is expected to reduce ethylene production. The expression level of *E4* and *E8* genes (ethylene-inducible genes) were also significantly reduced by the sound wave treatment (Fig. 5), probably due to the low concentrations of ethylene in sound-treated tomato fruit. These results well agree with the previous report on the strict dependence of those target genes with ethylene during fruit ripening (Kesanakurti et al., 2012 and Lincoln and Fischer, 1988).

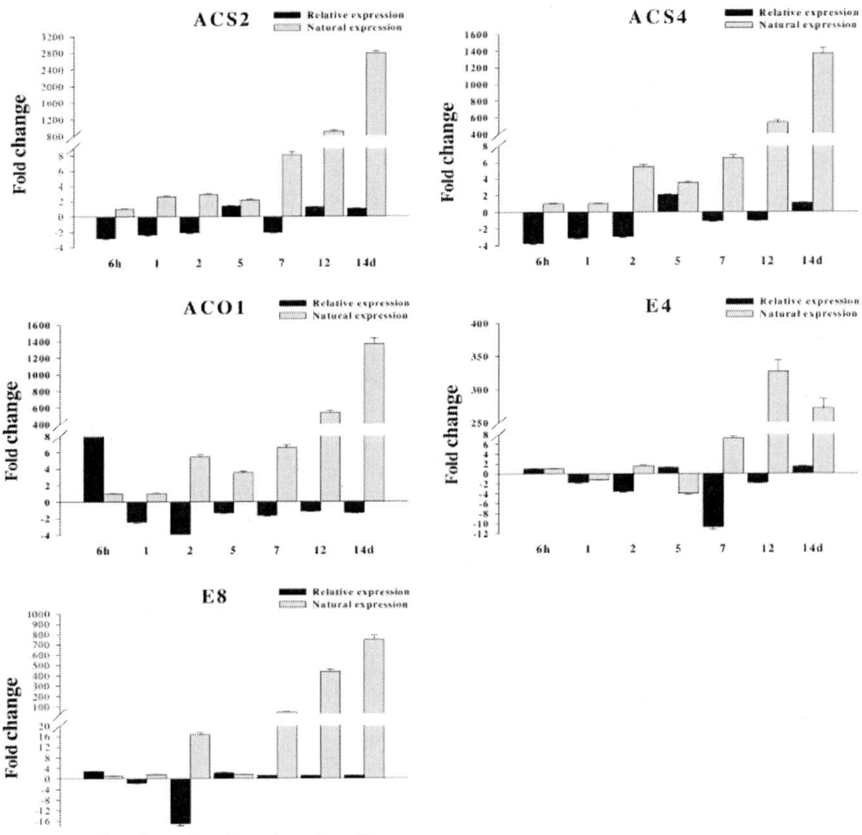

Figure 5. Expression of ethylene biosynthetic genes in tomato fruit after sound treatment.Expression of ethylene biosynthesis-related genes in tomato fruit at the indicated times (at 23 °C) after sound treatment (1 kHz) for 6 h, as determined by qRT-PCR. Natural expression is genes expression in non-treatment tomato fruit. Relative expression is value compared with natural expression in 1 kHz sound treatment tomato fruit. Error bars represent SD of three biological replicates.

3.5. Sound wave treatment triggers expression of ripening-related genes

RIN, LeHB-1, CNR, and *NOR* encode transcription factors that regulate the expression of several genes responsible for ethylene biosynthesis. *RIN* belongs to the MADS-box family of transcription factors, which are important regulators of vegetative growth, flowering time, and floral development (Vrebalov et al., 2002). *LeHB-1,* a class-I HD-Zip gene, encodes a transcription factor that binds to the promoters of ethylene biosynthesis-related genes that have recently been identified (Lin et al., 2008). In addition, the*TOMATO AGAMOUS-LIKE 1* (*TAGL1*) MADS box gene encodes a transcription factor that plays an important function in both fruit expansion and ripening. *TAGL1* promotes ripening by activating *ACS2*, and is reported to be a target of *RIN* (Itkin et al., 2009 and Vrebalov et al., 2009). Silencing of *TAGL1* or *RIN* delays ripening (Vrebalov et al., 2009, Giménez et al., 2010 and Pan et al., 2010). Recessive *rin* mutant tomatoes fail to ripen, as these mutants do not produce high levels of endogenous ethylene and do not respond to exogenous ethylene. *TAGL1* is expressed at early and later stages at the onset of ripening. Our results showed that the mRNA level of *RIN* decreased significantly as compared with the non-treated control fruit (Fig. 6). In fruit treated with sound waves, the expression level of *LeHB-1* was reduced after 1 day and 2 days, but increased after 5 days. Furthermore, *TAGL1* expression was reduced in the treated fruit, except between 7 and 12 days after sound wave treatment.

Also, *CNR* and *NOR* have been reported to regulate the expression of genes related to fruit ripening, including ethylene and carotenoid biosynthesis (Giovannoni, 2004). In this study, the transcript level of *CNR* decreased at 1–3 days after sound wave treatment, but increased at 7 days after treatment. The expression pattern of *CNR* was opposite in the sound-treated and non-treated fruit (Fig. 6). Meanwhile, the expression of *NOR* was very low in sound-treated fruit compared to the non-treated fruit throughout the 14-day observation period (Fig. 6). These results showed that sound wave treatment delayed tomato fruit ripening by inhibiting the expression of some of the most important ethylene biosynthetic and ripening regulatory genes.

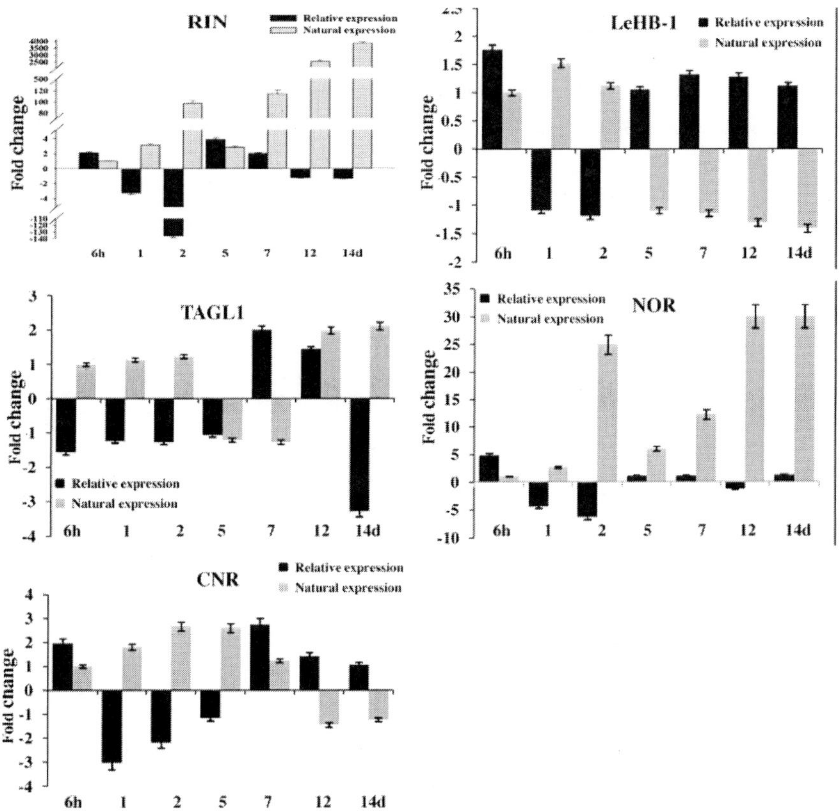

Figure 6. Expression of ripening-regulated genes in tomato fruit after sound treatment.Expression of ethylene response-related genes in tomato fruit at the indicated time points (at 23 °C) after sound treatment (1 kHz) for 6 h, as determined by qRT-PCR. Natural expression is genes expression in non-treatment tomato fruit. Relative expression is value compared with natural expression in 1 kHz sound treatment tomato fruit. Error bars represent SD of three biological replicates.

This study also showed that the mRNA levels of *RIN*, *ACS2,* and *ACS4* significantly increased after 7 days in non-treated fruit. However, the expression of these three genes was low, and remained low up to 14 days after sound wave treatment (Fig. 5 and Fig. 6).Martel et al. (2011) showed that *RIN* binds to the *ACS2* and *ACS4* promoter, but not to the*ACO1* promoter. *RIN* might be involved in reducing *ACS2* and *ACS4* expression. Due to the decreased expression of *RIN*, the expression of *ACS2* and *ACS4* may have been reduced, but the possibility cannot be ruled out that *ACS2* and *ACS4* expression is directly controlled by sound waves.

LeHB-1 has been reported to bind to the *LeACO1* promoter, which contains the putative HD protein binding sequence, and directly regulates a number of ripening-related genes (Lin et al., 2009). Manavella et al. (2006) showed that transgenic Arabidopsis plants harboring sunflower *Hahb-4* exhibited a remarkable delay in senescence and were less sensitive to ethylene. The expression of this gene suppresses the expression of genes associated with ethylene biosynthesis, such as *LeACO,* and with ethylene signaling, such as *ERF2* and *ERF5* (Manavella et al., 2006). Blume and Grierson (1997) reported that *LeHB-1* is highly expressed in mature green fruit and decreased at the breaker stage, whereas the expression of *LeACO1* increased in mature green fruit and accumulated during ripening. These results showed that the level of *LeHB-1* transcript declined in the green mature stage and increased after the breaker stage, whereas the expression level of *LeACO1* was higher in the early stage of ethylene biosynthesis and was maintained at low levels for up to 14 days after sound wave treatment, indicating that the expression pattern of these genes was affected by the sound wave treatment. As*LeHB-1* expression increased, the expression of *ACO1* was significantly reduced in sound wave-treated tomato fruit (Fig. 5 and Fig. 6).

TAGL1 is known to be involved in ripening in tomato (Itkin et al., 2009 and Vrebalov et al., 2009). *TAGL1* promotes ripening by activating *ACS2*, and is reported to be a target of*RIN* (Itkin et al., 2009 and Vrebalov et al., 2009). Silencing of *TAGL1* or *RIN* delays ripening (Vrebalov et al., 2009, Giménez et al., 2010 and Pan et al., 2010). Recessive*RIN* mutant tomatoes fail to ripen, as these mutants do not produce high levels of endogenous ethylene and do not respond to exogenous ethylene. *TAGL1* is expressed at early and later stages at the onset of ripening. These real-time RT-PCR results were similar to those reported by others (Busi et al., 2003 and Hileman et al., 2006) in non-treated fruit, but *TAGL1* expression was reduced during the early stage of ripening in tomato fruit treated with sound waves.

Every year, spoilage of postharvest agricultural products results in tremendous economic losses. The purpose of this study was to test the efficacy of delaying the ripening of tomato fruit by controlling the expression of genes related to ethylene biosynthesis through sound treatment. These results propose well that sound wave treatment delays tomato fruit ripening by regulating the expression of ethylene biosynthesis-related genes.

4. CONCLUSION

Findings in this study leads a conclusion that sound wave can be an external stimulus to delay the ripening of tomato fruit. This study investigated relevant molecular mechanisms through analysis of the expression of ethylene synthesis-related genes, controlling fruit ripening.

This study set the stage for further studies of sound wave-mediated responses in plants. Findings of this study indicate that acoustic biology can have viable applications in agriculture and may help to increase product quality and reduce losses. This study also provides the first indication that sound treatment can extend the ripening of tomato fruit. Since this area is still in the beginning stages, more detailed studies need to be performed in the future.

ACKNOWLEDGEMENT

This study was financially supported by NAAS Agenda Program (No. PJ0100252015 and PJ01049701) in Rural Development Administration, Jeonju, Republic of Korea.

REFERENCES

1. Alexander, L., Grierson, D., 2002. Ethylene biosynthesis and action in tomato: a model for climacteric fruit ripening. J. Exp. Bot. 53, 2039–2055. Barry, C.S., Giovannoni, J.J., 2007. Ethylene and fruit ripening. J. Plant Growth Regul. 26, 143–159.

2. Blume, B., Grierson, D., 1997. Expression of ACC oxidase promoter-GUS fusions in tomato and Nicotiana plumbaginifolia regulated by developmental and environmental stimuli. Plant J. 12 (4), 731–746.

3. Bochu, W., Hucheng, Z., Yiyao, L., Yi, J., Sakanishi, A., 2001. The effects of alternative stress on the cell membrane deformability of chrysanthemum callus cells. Colloids Surf. B 20, 321–325.

4. Bochu, W., Jiping, S., Biao, L., Jie, L., Chuanren, D., 2004. Sound wave stimulation triggers the content change of the endogenous hormone of the Chrysanthemum mature callus. Colloids Surf. B 37, 107–112.

5. Braam, J., 2004. In touch: plant response to mechanical stimuli. New Phytol. 165, 373–389.

6. Busi, M.V., Bustamante, C., D'Angelo, C., Hidalgo-Cuevas, M., Boggio, S.B., Valle, E.M., Zabaleta, E., 2003. MADS-box genes expressed during tomato seed and fruit development. Plant Mol. Biol. 52, 801–815.

7. Cara, B., Giovannoni, J.J., 2008. Molecular biology of ethylene during tomato fruit development and maturation. Plant Sci. 175, 106–113.

8. Giménez, E., Pineda, B., Capel, J., Antón, M.T., Atarés, A., Pérez-Martín, F., 2010. Functional analysis of the Arlequin mutant corroborates the essential role of the Arlequin/TAGL1 gene during reproductive development of tomato. PLoS One 5, e14427.

9. Giovannoni, J.J., 2004. Genetic regulation of fruit development and ripening. Plant Cell 16, S170–S180.

10. Hileman, L.C., Sundstrom, J.F., Litt, A., Chen, M., Shumba, T., Irish, V.F., 2006. Molecular and phylogenetic analyses of the MADS-box gene family in tomato. Mol. Biol. Evol. 23, 2245–2258.

11. Hoogstrate, S.W., van Bussel, L.J., Cristescu, S.M., Cator, E., Mariani, C., Vriezen, W.H., Rieu, I., 2014. Tomato ACS4 is

necessary for timely start of and progression through the climacteric phase of fruit ripening. Front Plant Sci. 5 (466), 1–7.

12. Itkin, M., Seybold, H., Breitel, D., Rogachev, L., Meir, S., Aharoni, A., 2009. TOMATO AGAMOUS-LIKE 1 is a component of the fruit ripening regulatory network. Plant J. 60, 1081–1095.

13. Jeong, M.J., Shim, C.K., Lee, J.O., Kwon, H.B., Kim, Y.H., Lee, S.K., Byun, M.O., Park, S.C., 2008. Plant gene reponses to frequency-specific sound signals. Mol. Breed. 21, 217–226.

14. Johnson, K.A., Sistrunk, M.L., Polisensky, D.H., Braam, J., 1998. Arabidopsis thaliana responses to mechanical stimulation do not required ETR1 or EIN2. Plant Physiol. 116, 643–649.

15. Johnson, P.R., Ecker, J.R., 1998. The ethylene gas signal transduction pathway: a molecular perspective. Annu. Rev. Genet. 32, 227–254.

16. Jomdecha, C., Prateepasen, A., 2011. Effects of pulse ultrasonic irradiation on the lag phase of Saccharomyces cerevisiae growth. Lett. Appl. Microbiol. 52, 62–69.

17. Katz, E., Lagunes, P.M., Riov, J., Weiss, D., Goldschmidt, E.E., 2004. Molecular and physiological evidence suggests the existence of a system II-like pathway of ethylene production in non-climacteric citrus fruit. Planta 219, 243–252.

18. Kesanakurti, D., Kolattukudy, P.E., Kirti, P.B., 2012. Fruit-specific overexpression of wound-induced tap1 under E8 promoter in tomato confers resistance to fungal pathogens at ripening stage. Physiol. Plant. 146, 136–148.

19. Lelièvre, J.-M., Latché, Jones, A., Bouzayen, B., Pech, M., J-C, 1997. Ethylene and fruit ripening. Physiol. Plant. 101, 727–739.

20. Lin, Z.F., Hong, Y., Yin, M., Li, C., Zhang, K., Grierson, D., 2008. A tomato HD-Zip homeobox protein, LeHB-1, plays animportant role in floral organogenesis and ripening. Plant J. 55, 301–310.

21. Lin, Z.F., Zhong, S.L., Grierson, D., 2009. Recent advances in ethylene research. J. Exp. Bot. 60, 3311–3336.

22. Lincoln, J.E., Fischer, R.L., 1988. Diverse mechanisms for the regulation of ethyleneinducible gene expression. Mol. Gen. Genet. 212, 71–75.

23. Lincoln, J.E., Campbell, A.D., Oetiker, J., Rottmann, W.H., Oeller, P.W., Shen, N.F., Theologis, A.,1993. Le-Acs4, a fruit ripening and wound-induced 1-aminocyclopropane-1-carboxylate synthase gene of tomato (Lycopersicon esculentum)- expression in E. coli, structural characterization, expression characteristics, and phylogenetic analysis. J. Biol. Chem. 268, 19422–19430.

24. Liu, M., Diretto, G., Pirrello, J., J-, P., Li, Z., Giuliano, G., Regad, F., Bouzayen, M., 2014. The chimeric repressor version of an ethylene response factor (ERF) family member, Sl-ERF.B3, shows contrasting effects on tomato fruit ripening. New Phytol. 203, 206–218.

25. Manavella, P.A., Arce, A.L., Dezar, C.A., Bitton, F., Renou, J.-P., Crespi, M., Chan, R.L., 2006. Cross-talk between ethylene and drought signaling pathways is mediated by the sunflower Hahb-4 transcription factor. Plant J. 48, 125–137.

26. Martel, C., Vrebalov, J., Tafelmeyer, P., Giovannoni, J.J., 2011. The tomato MADSBox transcription factor RIPENING INHIBITOR interacts with promoters involved in numerous ripening processes in a COLORLESS NONRIPENINGdependent manner. Plant Physiol. 157, 1568–1579.

27. Monshausen, G.B., Gilroy, S., 2009. Feeling green: mechanosensing in plants. Trends Cell Biol. 19, 228–235.

28. Pan, I.L., McQuinn, R., Giovannoni, J.J., Irish, V.F., 2010. Functional diversification of AGAMOUS lineage genes in regulating tomato flower and fruit development. J. Exp. Bot. 61, 1795–1806.

29. Takahashi, H., Suge, H., Kato, T., 1992. Growth promotion by vibration at 50 Hz in rice and cucumber seedlings. Plant Cell Physiol. 32, 729–732.

30. Vrebalov, J., Ruezinsky, D., Padmanabhan, V., White, R., Medrano, D., Drake, R., Schuch, W., Giovannoni, J., 2002. A MADS-box gene necessary for fruit ripening at the tomato ripening inhibitor (rin) locus. Science 296, 343–346.

31. Vrebalov, J., Pan, I.L., Arroyo, A.J.M., McQuinn, R., Chung, M., Poole, M., 2009. Fleshy fruit expansion and ripening are regulated by the tomato SHATTERPROOF gene TAGL1. Plant Cell 21, 3041–3062.

32. Weinberger, P., Measures, M., 1978. Effects of the intensity of audible sound on the growth and development of Rideau winter wheat. Can. J. Bot. 57, 1036–1039. Yang, S.F., Hoffman, N.E., 1984. Ethylene biosynthesis and its regulation in higherplants. Annu. Rev. Plant Physiol. 35, 155–189.

33. Yokotani, N., Tamura, S., Nakano, R., Inaba, A., McGlasson, W.B., Kubo, Y., 2004. Comparison of ethylene-wound-induced responses in fruit of wild-type, rin and nor tomatoes. Postharvest Biol. Technol. 32, 247–252.

13

MORPHOLOGICAL AND MOLECULAR CHARACTERIZATION OF ETHYLENE BINDING INHIBITION IN CARNATIONS

B.C. Ina, J. Strablea,b, B.M. Binder c, T.G. Falbel a, S.E. Pattersona

[a]Department of Horticulture, University of Wisconsin-Madison, Madison, WI 53706, USA

[b]Department of Genetics, Development and Cell Biology, Iowa State University, Ames, IA 50011, USA

[c]Department of Biochemistry, Cellular, and Molecular Biology, University of Tennessee, Knoxville, TN 37996, USA

ABSTRACT

Ethylene perception by ethylene receptors can be suppressed by ethylene antagonists, such as 1-methylcyclopropene (1-MCP). 1-MCP binds to ethylene receptors blocking ethylene binding, and thereby represses ethylene responses, including flower senescence and petal abscission. Despite its antagonistic propensity, plants treated with 1-MCP often regain sensitivity to ethylene, suggesting that ethylene receptors are synthesized de novo post-treatment. To investigate this observation, we determined the relationship between the mRNA levels of ethylene biosynthesis and ethylene receptor genes and the degree of ethylene sensitivity of carnation flowers after ethylene and 1-MCP treatments. Flowers treated with a single application of 1-MCP lost the inhibitory effect in ethylene production after 7 d of the treatment. In contrast, multiple treatments with 1-MCP completely suppressed several ethylene biosynthesis genes and ethylene production throughout the experiment. Multiple 1-MCP treatments increased vase life by almost three-fold relative to control flowers. Eventually ethylene-independent browning and desiccation on the edges of petals contributed to loss of vase life. We observed that the transcript levels of the ethylene receptor genes, *DcETR1* and *DcERS1*, decreased with flower development, but increased at the onset of floral senescence. Our results suggest that ethylene receptors are continually synthesized during later stages of floral development. While ethylene responsiveness is temporarily blocked by 1-MCP treatment, we predict that new receptors synthesized days after 1-MCP treatment can bind ethylene. This leads to the observed recovery of ethylene sensitivity in the flowers. In addition, we predict that the degradation of ethylene receptors that is stimulated by ethylene binding is prevented by successive treatments with 1-MCP prior to recovery of ethylene sensitivity. This study improves our understanding about the uses and effectiveness of 1-MCP and will facilitate the development and improvement of postharvest treatments for multiple crops.

Keywords: Carnation; Ethylene biosynthesis; Ethylene receptor; Flower senescence; 1-MCP; Cut flowers

1. INTRODUCTION

Cut flower longevity has been a concern of growers for decades. Although there are numerous practices and treatments to prolong the vase life of cut flowers, the process of flower senescence is still poorly understood. The plant hormone ethylene regulates multiple aspects of plant growth and development, including seed germination, abscission, flower senescence, and fruit ripening (Abeles et al., 1992 and Reid and Wu, 1992). Ethylene is synthesized throughout the plant via a well-characterized biosynthetic pathway (Yang and Hoffman, 1984 and Kende, 1993).

In many flowers, the onset of flower senescence is associated with a climacteric increase in ethylene biosynthesis. Once the autocatalytic synthesis of ethylene is initiated, the senescence process is considered irreversible in ethylene-sensitive flowers such as carnations (Halevy and Mayak, 1981 and Yang and Hoffman, 1984). Ethylene perception by ethylene receptors is an indispensable requirement to commence and sustain the ethylene-mediated senescence program (Borochov and Woodson, 1989). When ethylene is perceived by the receptors, the ethylene signal is sent through a sequence of biochemical events that regulate the expression of ethylene responsive genes leading to ethylene production and ultimately flower senescence. The ability to perceive or respond to ethylene is most likely mediated by changes in ethylene signaling during flower development (Halevy and Mayak, 1981, Woodson and Lawton, 1988 and Verlinden et al., 2002).

From multiple genetic studies in the model plant *Arabidopsis thaliana*, a family of ethylene receptors (ETR1, ETR2, ERS1, ERS2 and EIN4), as well as several downstream genes (CTR1, EIN2, EIN3, EIL1, and ERF1) have been discovered that act in a signal transduction cascade (Schaller, 2012). Ethylene perception by the family of receptors is thought to inactivate the antagonist CTR1, a Raf-like MAP kinase kinase kinase (MAPKKK), thereby repressing the negative regulation of these receptors at the cell membrane (Chang et al., 1993, Kieber et al., 1993, Hua and Meyerowitz, 1998, Hua et al., 1998 and Sakai et al., 1998).

Historically, it has been shown that the ethylene receptor isoforms have ethylene binding sites with similar binding affinities for ethylene (Schaller and Bleecker, 1995, Hall et al., 2000 and O'Malley et al., 2005). Ethylene binding and action can be suppressed by the potent inhibitor 1-methylcyclopropene (1-MCP). Interestingly, 1-MCP has been shown to bind competitively to the ethylene receptors with a greater affinity for the receptors than that of ethylene (Sisler et al., 1986, Sisler et al., 1990 and Hall et al., 2000). Therefore, it has been proposed that 1-MCP can block the ethylene binding to the receptor and inhibit down-stream components in ethylene signal transduction. Consequently, it was found that 1-MCP works effectively in many horticultural plants to prolong their postharvest longevity such as carnation (Sisler et al., 1996), *Alstroemeria*, *Kalanchoe* (Serek et al., 1994 and Serek et al., 1995), *Pelargonium* (Cameron and Reid, 2001), banana, avocado and apple (Jiang et al., 1999, Watkins et al., 2000, Blankenship and Dole, 2003 and Schotsmans et al., 2009). In addition, it has been shown that 1-MCP is effective at very low concentration (10–50 nL L^{-1} for 6 h) in carnation flowers (Serek et al., 1995,Sisler et al., 1996 and Reid and Çelikel, 2008).

1-MCP is hypothesized to be bound to ethylene receptors irreversibly *in planta*; and thus, plants treated with 1-MCP should not respond to ethylene (Serek et al., 1994, Sisler et al., 1996, Sisler and Serek, 2003 and Binder et al., 2004a). Despite this irreversible binding to the receptors, plants treated with 1-MCP often regain sensitivity to ethylene (i.e., recover) post-treatment. However, plants that undergo successive 1-MCP treatments prior to their recovery of ethylene sensitivity have prolonged repression of ethylene responses (Cameron and Reid, 2001 and Ekman et al., 2004).

Based on an irreversible antagonist model for 1-MCP action, it has been proposed that the recovery of ethylene sensitivity in plants after 1-MCP treatment is the consequence of synthesis of new ethylene receptors during development and senescence of the flowers (Binder and Bleecker, 2003 and Schotsmans et al., 2009). However, to our knowledge, no data directly supports this hypothesis and it is not understood why some plants respond differently than others. In this study, we set out to test the hypothesis that ethylene receptors are synthesized de novo, and that this accounts for the ability of the carnation plants to regain ethylene sensitivity post-treatment with 1-MCP.

Carnation (*Dianthus caryophyllus* L.) flowers were selected for this study due to their high sensitivity and rapid response to gaseous regulators such as ethylene and 1-MCP. Three ethylene receptor genes have been identified in carnation plants, *DcETR1*, *DcERS1* and*DcERS2* that are orthologous to the ethylene receptors of *Arabidopsis* (Charng et al., 1997, Shibuya et al., 1998 and Nagata et al., 2000). We took a molecular approach to explore how plants regain the sensitivity to ethylene at the transcript level by monitoring the relationship between mRNA levels of ethylene biosynthesis genes and receptor genes, and the sensitivity of flowers to ethylene at various stages of development after single or multiple 1-MCP treatments. Floral morphology was also observed before and after treatments to determine the effects of 1-MCP and ethylene on carnation flowers.

2. MATERIALS AND METHODS

2.1. Plant materials

Rooted cuttings of carnation plants (*Dianthus caryophyllus* L. cv. Moonstone) were planted in 15 cm plastic pots with growing medium (Metro-Mix Special Blend, SUNGRO Horticulture Distribution Inc., Bellevue, WA). The plants were grown on benches in a greenhouse at 22/16 °C day/night temperature and flood-irrigated with half strength Hoagland nutrient solution. Supplementary lighting (220 $\mu mol\ m^{-2}\ s^{-1}$ PPFD at plant level) was provided by high-pressure sodium lamps (Philips 600 W Master GreenPower) from 05:00 until 21:00 to assure a photoperiod of 16 h. These lamps were turned on and off automatically by a light sensor located inside the ceiling of the greenhouse that operated when the intensity of natural lighting was less than or exceeded 350 $\mu mol\ m^{-2}\ s^{-1}$. Plants were pinched 4 weeks after planting. Carnation flowers were harvested when the outer petals were just horizontal and immediately placed in tap water for all experiments.

2.2. Treatment with ethylene and 1-MCP

Flowers were trimmed to 5 cm stem length, and placed in 20 mL glass vials containing distilled water. The flowers were subsequently treated singly with air, 5 $\mu L\ L^{-1}$ ethylene, 100 $nL\ L^{-1}$ 1-MCP for 6 h; or repeated 1-MCP treatments every other day. Air treated flowers were used as controls. Control and treated flowers were enclosed in dark treatment chambers (117 L) with air circulation by a small fan. Control flowers were maintained in the chamber with normal atmospheric air. For ethylene treatment, 5.85 mL of 10% ethylene was injected into the chamber to give a final concentration of 5 μL L^{-1}. The ethylene action inhibitor, 1-MCP was applied via SmartFresh (AgroFresh Inc., Philadelphia, PA) tablets placed in the treatment chambers at room temperature. To generate 100 nL L^{-1} 1-MCP gas, a gray research tablet (0.026 mg 1-MCP) and an activator tablet was added to a beaker containing 18 mL of activator solution and was placed in the chamber. The cut flowers were transferred to the chamber along with small fan to provide air circulation during the treatments (ethylene or air) and then the chamber was tightly sealed for 6 h. After the treatments, the flowers were held in environmental controlled room at 20 ± 2 °C, 40 ± 5% relative humidity and about 50 μmol m^{-2} s^{-1} of fluorescent lighting for 12 h.

2.3. Ethylene measurements

For ethylene measurements, three carnation flowers in each treatment were selected at four distinct stages of flower development: day 1, flowers beginning to open; day 3, flowers completely opened; day 6, flowers showing onset of petal in-rolling; and day 7, flowers with the petal in-rolled. Individual flowers were placed in a 500 mL glass jar with a rubber septum and were enclosed for 1 h at room temperature. A 1 mL gas sample was collected with a gas-tight hypodermic syringe through the septum and analyzed for ethylene by a gas chromatograph (Model 8500, PerkinElmer Corp., Norwalk, CT) equipped with an aluminum column and flame ionization detector. Each experiment was repeated at least three times.

2.4. cDNA synthesis and semi-quantitative RT-PCR

Receptacles, petal claw and petal blade were used for the gene expression analysis since the floral organs show a distinct differences in the responsiveness to ethylene (Overbeek and Woltering, 1990 and Drory et al., 1993). After the assay for ethylene production, petals and the receptacle were detached from the flowers and the petals were divided into upper (petal blade) and basal (petal claw) portions (Mor et al., 1985). For tissue-specific expression profiles, petals, ovaries, receptacles, styles, stamens, and sepals were also detached from fully opened flowers. Immediately after detachment, they were frozen in liquid nitrogen and stored at −80 °C for RNA isolation. Total RNA was isolated from the various floral organs using Trizol method according to the manufacturer's procedure with slight modifications. Approximately 300 mg of tissues were ground in the presence of liquid nitrogen to a fine powder using a pre-chilled mortar and pestle and were homogenized in 1 mL Trizol Reagent (Invitrogen, Carlsbad, CA). The sample in Trizol was separated into aqueous and hydrophobic phases with 200 μL chloroform by centrifugation. The RNA was precipitated from the aqueous phase with 50 μL of 5 M NaCl and 0.5 mL isopropanol by centrifugation, washed with 80% ethanol, and resuspended in RNase-free H_2O. Total RNA was quantified with NanoDrop DN-1000 spectrophotometer (NanoDrop Technologies Inc., Wilmington, DE). RNA samples were treated with RNase-free DNase prior to RT-PCR and first-strand cDNA was synthesized from 2 μg of total RNA with 1 μg of oligo $(dT)_{18}$ primer, dNTPs, RNase inhibitor, buffer, and M-MLV reverse transcriptase in a final volume of 25 μL according to the manufacturer's instructions (Promega, WI). Reverse transcription was performed in a PTC-200 PCR machine (MJ Research Inc., MA) with the following temperature parameters: 15 min at 70 °C followed by 1 h at 42 °C.

To determine the expression patterns of ethylene biosynthesis and receptor genes during flower senescence, a semi-quantitative RT-PCR analysis of the transcript levels was performed using gene-specific primers. These experiments were performed with three independent biological replicates. Based on database published sequences, a pair of specific primers was designed for *DcACS1*, *DcACO1*, *DcETR1*, *DcERS1* and *DcERS2*genes and synthesized by Integrated DNA Technologies (Coralville, IA). Carnation actin (*DcACT1*) was used as an internal control. Primer pairs that were used for the semi-quantitative PCR analyses are listed in Table 1. For PCR reactions, 1 μL of cDNA was used as a template with 2 μL of PCR buffer, 0.2 μL of Taq polymerase, dNTPs, and forward and reverse primers in a final volume of 20 μL. PCR amplification was performed with the following temperature parameters: 95 °C for 3 min and then cycled 35 (*DcERS2)*or 40 times (all the other genes) at 95 °C for 20 s, 60 °C for 30 s, and 72 °C for 40 s, with a final extension of 7 min at 72 °C. The PCR products were then analyzed by 1% (w/v) agarose gel and stained with ethidium bromide. The gels were visualized under UV light and the images were taken using a gel documentation system (Gel Doc XR+, Bio-Rad, Hercules, CA) and quantified with ImageJ 1.43u (National Institutes of Health, USA). The relative level of the band was showed as the absolute integrated absorbency normalized to the relative actin band.

Table 1. Gene-specific primers used for RT-PCR amplification of cDNA fragments.

Gene (accession number)	Forward primer	Reverse primer	Cycles
DcACS1 (M66619)	5′-AAAATCGCTACCAACGATCG-3′	5′-TGCGGCATAATAAGGAGACG-3′	40
DcACO1 (AB042320)	5′-TCACAACTGGGGATTCTTCC-3′	5′-CATCTGTCTGCGCTATCACG-3′	40
DcETR1 (AB035806.1)	5′-CACACTCCGTCACCAATATCC-3′	5′-TCCATAAGCTCTTCCTGTGC-3′	40
DcERS1 (AF016250.1)	5′-CGTCTCTGCTGTAGGTGACG-3′	5′-CAAATTGCGAGTCCAAGTCC-3′	40
DcERS2 (AF034770.1)	5′-AGAGTGACGGTCATGACAAGG-3′	5′-CCTAACCGCGCTTAACTCC-3′	35
DcACT1 (AY007315)	5′-GGCCGTTCTCTCTCTGTACG-3′	5′-GTCCATCAGGCATTTCATAGC-3′	35

3. RESULTS

3.1. Suppression of ethylene biosynthesis in floral organs by 1-MCP treatment

To understand the relationship between ethylene biosynthesis and flower senescence, the expression of ethylene biosynthesis genes, ethylene production, and flower longevity were determined at four stages of flower development after treatment with ethylene, a single application of 1-MCP, or multiple applications of 1-MCP. The expression levels of ethylene biosynthesis genes *DcACS1* and *DcACO1* were monitored during flower development and senescence (Fig. 1). The transcript levels of *DcACS1* and *DcACO1* in control flowers were very low or undetectable until day 3 and increased significantly in all floral organs during late senescent stages (days 6 and 7). In response to ethylene treatment, initial transcript levels of *DcACO1* increased significantly in all tissues (Fig. 1B) while changes in levels of *DcACS1* were not discernable (Fig. 1A). These findings enhance previous observations on spatial and temporal specificity of ethylene biosyntheis genes in carnation petals. Previous studies reported differences in responsiveness of these ethylene biosynthesis genes indicating that the expression levels of *DcACS1* are similar or more abundant than *DcACO1* during natural senescence and that *DcACO1* is rapidly induced by ethylene treatment (Woodson et al., 1992 and ten

Have and Woltering, 1997). We also observed that a single treatment with 1-MCP completely inhibited the changes in transcript levels of *DcACS1* in specific tissues including petal claw and petal blade. However, there was no measurable effect on*DcACS1* expression in the receptacle.

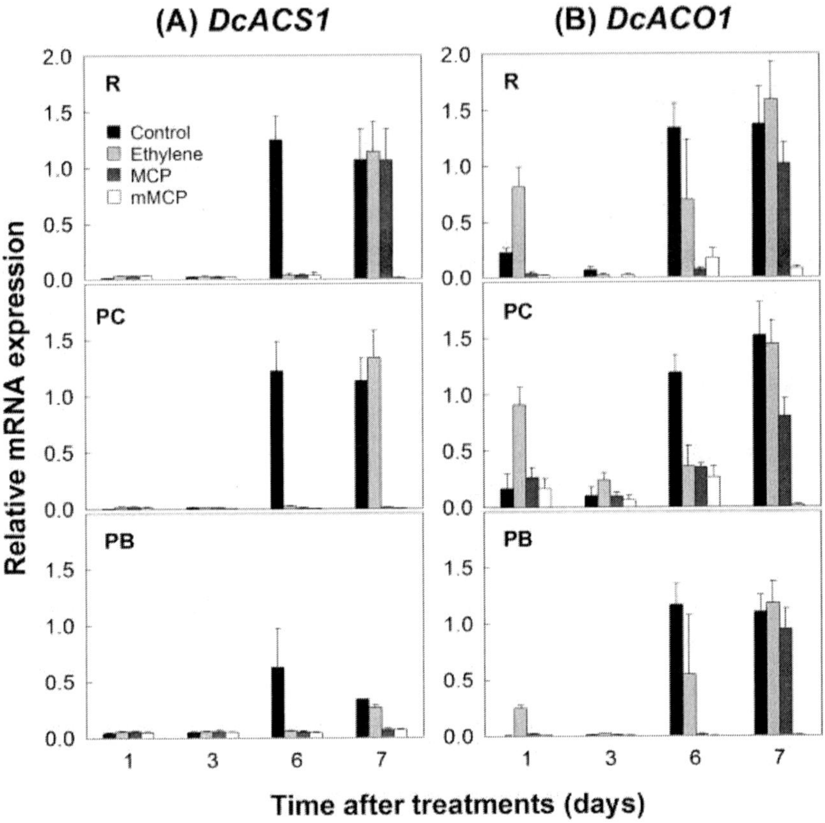

Figure 1. Expression patterns of *DcACS1 and DcACO1* genes in floral organs of carnations. Flowers were harvested at the fully opened stage and immediately treated singly with ethylene (5 μL L^{-1}); 1-MCP (100 nL L^{-1}) for 6 h; or multiple 1-MCP treatments every other day. Total RNA was isolated from the receptacle (R), petal claw (PC), and petal blade (PB) immediately after the determination of ethylene production in whole flowers. Semi-quantitative RT-PCR analysis was performed to determine the mRNA level of *DcACS1* and*DcACO1*. The relative expression level was normalized to actin (*DcACT1*). Control, Air only; MCP, single 1-MCP treatment; mMCP, multiple 1-MCP treatments at day 1, day 3, day 5 and day 7. Data represents the mean ± SE of three replicates.

We determined that ethylene production in whole flowers primarily mirrored the expression patterns of the ethylene biosynthesis genes *DcACS1* and *DcACO1* (Fig. 1 and Fig. 2). Production in untreated control flowers showed a dramatic increase at day 6, followed by a decrease at day 7. Flowers treated with ethylene produced 30% higher levels of ethylene. In contrast, flowers treated with a single application of 1-MCP displayed a delayed increase in ethylene production that was not measurable until day 7. These plants produced approximately 50% lower levels of ethylene than control plants. Flowers treated with multiple applications of 1-MCP had no detectable ethylene production.

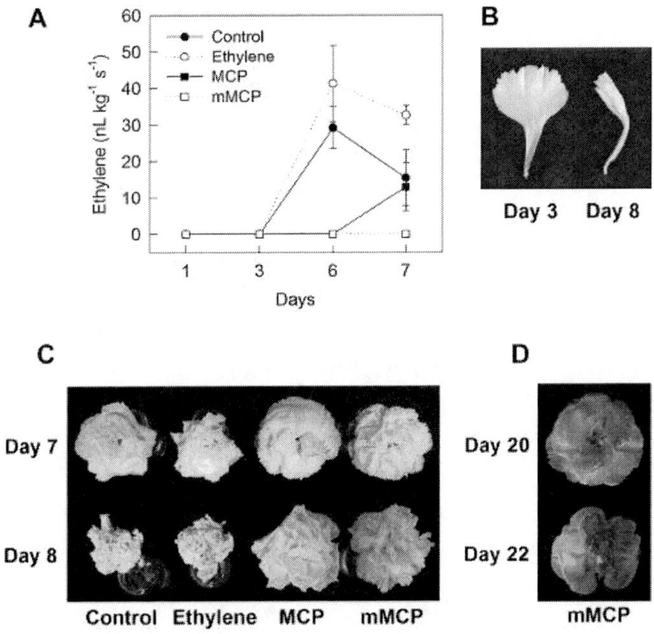

Figure 2. Effects of ethylene and 1-MCP treatment on ethylene production and flower senescence during vase life in carnation cv. Moonstone. (A) Changes in ethylene production of whole flowers; (B) carnation petal from a fully open flower (Day 3) and senescent flower showing inrolling (Day 8); (C) flower senescence on day 7 and day 8 after treatments with ethylene or 1-MCP; (D) flower senescence on day 20 and day 22 in the flowers treated multiple times with 1-MCP. Flowers were harvested at the fully opened stage and immediately treated singly with ethylene (5 μL L^{-1}); 1-MCP (100 nL L^{-1}) for 6 h; or multiple 1-MCP treatments every other day. Control, Air only; MCP, 1-MCP single; mMCP, multiple 1-MCP treatments at day 1, day 3, day 5 and day 7. Data represents the mean ± SE of three replicates.

Flower longevity was strongly correlated with ethylene production patterns. Both control and ethylene-treated flowers demonstrated obvious signs of senescence by day 7 with more pronounced symptoms in the ethylene treated flowers. Specific morphological changes included petal inrolling and wilting, and these flowers completely ended their vase life on day 8 due to these changes (Fig. 2B and C). Vase life of the flowers after a single 1-MCP treatment was prolonged by about 2 d relative to control flowers. When the flowers were treated with multiple applications of 1-MCP, vase life was extended by 14 d compared to control flowers (Fig. 2C). The flowers treated with multiple applications of 1-MCP did not show petal inrolling during the vase period. However, they eventually lost their vase life after 20 d due to browning or desiccation around the petal margins (Fig. 2D); a senescence symptom that occurs in many ethylene-insensitive floral organs (Tanase et al., 2008).

3.2. Tissue-specific expression profiles of ethylene biosynthesis and receptor genes

In order to understand ethylene responses in carnation flowers more fully, we also looked at the relationship of tissue specific expression of ethylene biosynthesis and receptor genes in the different floral organ tissues using RT-PCR. The expression profiles of ethylene biosynthesis and receptor genes revealed tissue-specific signatures in floral organs of carnations at the fully opened stage (Fig. 3). The ethylene biosynthesis gene*DcACS1* was not detected in most of the tissues with the exception of weak expression in the ovary. Alternately, *DcACO1* transcript was abundant in the receptacle, ovary, stigma and stamen. However, *DcACO1* expressed very weakly in the petal claw, and was not detected in the petal blade of flowers at the fully opened stage. Transcripts for all three ethylene receptor genes were found in all tissues studied; although, the patterns of expression varied for each receptor isoform. *DcETR1* transcript levels were generally lower than the other isoforms. Alternatively, *DcERS1* transcripts were more abundant in the receptacle, ovary and style, and less abundant in the petal and stamen. However,*DcERS2* transcripts generally showed equal abundance in all floral organs except the petal claw where there was less transcript (Fig. 3).

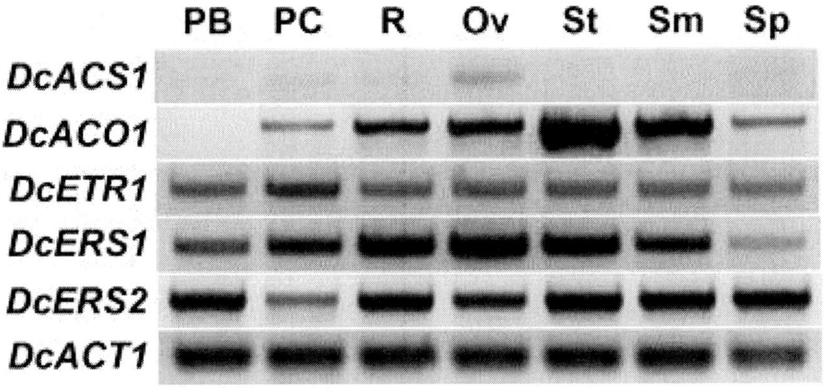

Figure 3. Tissue-specific expression profiles of *DcACS1*, *DcACO1*, *DcETR1*, *DcERS1* and *DcERS2* genes in carnation flowers. Expression analysis was performed on various floral organs of carnations at the fully opened stage using RT-PCR. PB, petal blade; PC, petal claw; R, receptacle; Ov, ovary; St, style; Sm, stamen; Sp, sepal.

3.3. Expression pattern of ethylene receptor genes during development and senescence of flowers

To further assess the role of ethylene receptor genes in ethylene sensitivity, the expression patterns of the three ethylene receptor genes, *DcETR1*, *DcERS1* and*DcERS2* were determined in the receptacle, petal claw and petal blade at various stages of flower development after treatments with ethylene or 1-MCP (Fig. 4). The expression of *DcETR1* in air control treatments was fairly constitutive in all tissues with a slight initial decrease in expression that might have been in response to recovery from harvest. We observed that *DcETR1* expression in the petal claw and petal blade was only slightly affected by either ethylene or 1-MCP treatments when compared with the controls. However, in the receptacle, *DcETR1* transcript accumulation was affected by both ethylene and 1-MCP treatments throughout floral development.

Subsequent to ethylene treatment, *DcETR1* transcript levels in the receptacle increased more than 2-fold relative to control at the later stages of flower development. While *DcETR1* expression was markedly increased by day 7 after ethylene treatment, its expression was suppressed by multiple applications of 1-MCP. We also observed slight increases in *DcETR1* transcript at days 6 or 7 in the receptacle for all treatments except the multiple 1-MCP treatments (Fig. 4A).

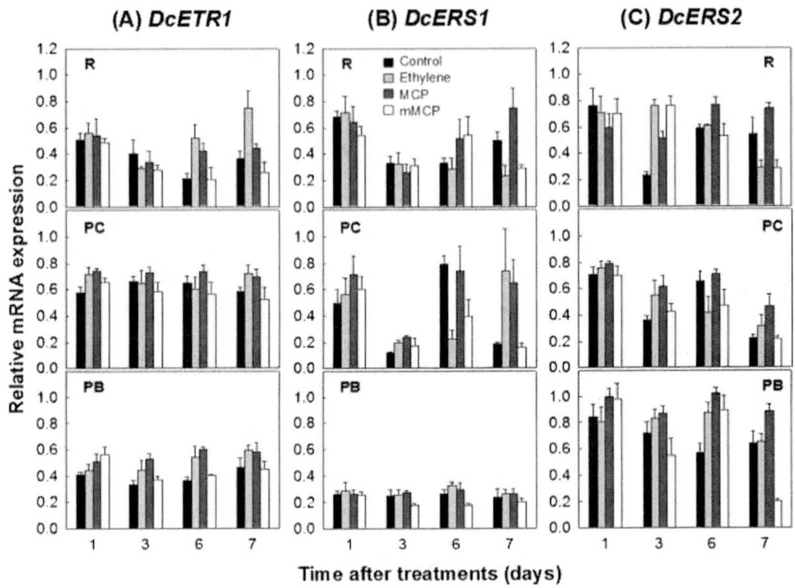

Figure 4. Expression patterns of *DcETR1, DcERS1* and *DCERS2* genes in floral organs of carnations. Flowers were harvested at the fully opened stage and immediately treated singly with ethylene (5 μL L^{-1}); 1-MCP (100 nL L^{-1}) for 6 h; or multiple 1-MCP treatments every other day. Total RNA was isolated from receptacle (R), petal claw (PC), and petal blade (PB) immediately after the determination of ethylene production in whole flowers as shown in Fig. 1. Semi-quantitative RT-PCR analysis was performed to determine the mRNA level of *DcETR1, DcERS1* and *DCERS2*. The relative expression level was normalized to actin (*DcACT1*). Control, Air only; MCP, single 1-MCP treatment; mMCP, multiple 1-MCP treatments at day 1, day 3, day 5 and day 7. Data represents the mean ± SE of three replicates.

In general, the expression level of *DcERS1* mRNA dropped rapidly at the fully open stage of flower development on day 3 and then showed increasing levels toward the late senescent stages in both receptacle and petal claw. Control tissues showed a consistently high level of *DcERS1* expression in petal blade throughout flower development and senescence. Following ethylene treatment, *DcERS1* transcript level increased considerably by day 7 in petal claw, but remained lower than untreated control in receptacle throughout flower development. In petal blades, *DcERS1 expression* was not clearly affected by ethylene treatment compared to control. Curiously, single applications of 1-MCP had no measurable effect on *DcERS1* expression in all floral organs. Overall, multiple treatments with 1-MCP suppressed changes in *DcERS1*transcript levels at the later stages of floral development. The mRNA levels were lowered by multiple applications of 1-MCP compared to control and ethylene-treated tissues. This was especially apparent in the petal tissues (Fig. 4B).

The mRNA levels of *DcERS2* in control plants initially showed higher accumulation and subsequently a rapid decrease in fully opened flowers. The *DcERS2* transcript in the receptacle increased in association with flower senescence at the later stages. In response to ethylene, *DcERS2* expression was elevated in receptacle tissues at the fully opened stage of flower development and reduced at later stages. However, in petal blade tissues, the expression patterns of *DcERS2* appeared unaffected by ethylene when compared with the control. *DcERS2* expression in 1-MCP single-treated tissues showed similar patterns with that in control tissues in receptacle. *DcERS2* expression decreased on day 3 and then increased at the late senescent stages. However, multiple treatments with 1-MCP resulted in decreased levels of expression in petal blade and receptacle compared to control at the later stage on day 7 (Fig. 4C).

These transcript data indicate that both *DcETR1* and *DcERS1* expression is more significantly changed than *DcERS2* expression in response to ethylene treatment. Overall, the expression levels of receptor genes was enhanced by ethylene treatment, and inhibited by multiple treatments with 1-MCP, indicating that higher levels of expression correlate with floral senescence.

3.4. Changes in floral morphology and petal turgidity

To understand the relationship between the expression patterns of ethylene receptor genes, flower sensitivity to ethylene, and morphological changes, the changes in petal turgidity were investigated throughout flower development and senescence. Petal turgidity was observed in carnation flowers after all sepals were removed from the flowers to exclude the influence of calyx. Control flowers maintained turgid petals until the fully open flower stage of development (day 3) with obvious wilting at day 6, and complete loss of turgidity and inrolling on day 7. While limited morphological differences in petals were observed between control and ethylene-treated flowers, ethylene treatment caused the flowers to lose petal turgidity slightly earlier than the control flowers. Flowers that were treated with 1-MCP (both singly and multiple times) lacked inrolling throughout the experiment indicating reduced loss of turgidity (Fig. 5).

Figure 5. Changes in petal turgidity in carnation flowers during vase life. Flowers were harvested at the fully opened stage and immediately treated singly with ethylene (5 μL L^{-1}); 1-MCP (100 nL L^{-1}) for 6 h; or 1-MCP multiple treatments every the other day. Photographs were taken after removal of sepals from the flowers to exclude the influence of sepal. Control, Air only; MCP, single 1-MCP treatment; mMCP, multiple 1-MCP treatments at day 1, day 3, day 5 and day 7.

4. DISCUSSION

Ethylene perception by ethylene receptors can be suppressed by the competitive inhibitor 1-MCP. However, the inhibitory effects of 1-MCP on ethylene responses are often transient (Cameron and Reid, 2001 and Ekman et al., 2004). To better understand the antagonistic effects of 1-MCP, we compared the levels of specific genes, ethylene production and vase life in cut carnations after treatment with ethylene versus single and multiple treatments with 1-MCP. Single treatment of cut carnation flowers with ethylene resulted in measurable increases in autocatalytic ethylene production. This was paralleled by increases in the transcript levels of specific ethylene biosynthesis genes. In contrast, single treatments with 1-MCP resulted in delay in the onset of ethylene production and decreases in overall levels of ethylene production. Multiple 1-MCP treatments blocked increases in ethylene production. Overall, analysis of transcript levels of ethylene biosynthesis genes was consistent with these results.

We also examined the effects of 1-MCP on the vase life of cut carnations. Flowers treated with a single 1-MCP application had prolonged vase life characterized by a delay in petal inrolling. In contrast, flowers treated with multiple applications of 1-MCP, did not show petal inrolling during the vase period, but eventually lost their vase life after 20 d due to browning or desiccation around the petal margins. Interestingly, this is a common senescence symptom that occurs in ethylene-insensitive floral organs suggesting that the multiple 1-MCP treatments are completely blocking ethylene perception. These results are consistent with previous works showing that re-treatment with 1-MCP prior to regaining ethylene sensitivity effectively prevented the ethylene responses in*Pelargoniums*, waxflowers and pears (Cameron and Reid, 2001, Ekman et al., 2004 and Macnish et al., 2004). Together, these results indicate that cut carnations can regain sensitivity to ethylene after a single 1-MCP treatment, but that the effects of ethylene are completely blocked by multiple treatments with 1-MCP.

We also observed tissue-specific differences in gene expression and regulation. In higher plants, the ethylene biosynthetic pathway involves the conversion of S-adenosylmethionine (SAM) to 1-aminocyclopropane-1-carboxylic acid (ACC) by ACC synthase (ACS), and

ACC is then converted to ethylene by ACC oxidase (ACO). ACS and ACO are generally regarded as rate-limiting enzymes in ethylene biosynthesis (Yang and Hoffman, 1984 and Kende, 1993). Presently, three ACS genes (*DcACS1, DcACS2*and *DcACS3*) and one ACO gene (*DcACO1*) have been identified in carnation (Park et al., 1992 and Henskens et al., 1994). *DcACS1* (*CARACC3*) has been shown to have a role in autocatalytic ethylene production in carnation flowers (ten Have and Woltering, 1997, Jones and Woodson, 1999, Jones, 2003 and Harada et al., 2011). In our examination of the transcript levels of ethylene biosynthesis genes in specific floral organ tissues we found that there were differences in the ethylene responsiveness of specific transcripts. Consistent with prior studies (Jones, 2003), we found that expression of the*DcACS1* and *DcACO1* transcripts were up-regulated slightly by ethylene and down-regulated by 1-MCP. A single treatment with 1-MCP prevented *DcACS1* transcript accumulation in petal claw and petal blade throughout the experimental period, but did not inhibit its expression levels in the receptacle at later stages pointing to key tissue-specific differences in the regulation of this gene. Interestingly, *DcACO1* transcript was transiently enhanced by exogenous ethylene on day 1 but did not result in increased ethylene production.

Transcripts for the receptors also showed interesting differences between the floral organs and the receptor isoforms. For instance, the transcript levels of *DcERS1* and*DcERS2* were much more abundant than that of *DcETR1* in the floral organ tissues of newly opened flowers suggesting these receptor isoforms have a larger role at this developmental stage. This is similar to the pattern observed in *Arabidopsis* where the*ETR1* and *ERS1* transcripts comprise more than 80% of total transcript levels of ethylene receptors (Binder et al., 2004b and O'Malley et al., 2005) and these receptor isoforms have a more pronounced role in ethylene signaling (Wang et al., 2003).

The accumulation of these transcripts was also influenced by ethylene and 1-MCP at a tissue-specific level. Especially, in receptacles, *DcETR1* transcript was significantly affected by both ethylene and multiple application of 1-MCP at the later stage, while in the petal tissues, it was only slightly influenced by either ethylene or 1-MCP. The difference in the gene expression of the receptors could be due to the different responsiveness to ethylene (and perhaps to 1-MCP) between the floral organs since the receptacles respond rapidly to ethylene by induction of *DcACS1* and *DcACO1* transcripts relative to petal tissues (Overbeek and Woltering, 1990 and Drory et al., 1993). Previous studies also showed that change in the receptor levels is associated with autocatalytic ethylene synthesis, which is mediated by the members of ACS family (Riov and Yang, 1982, Barry and Giovannoni, 2007 and Kevany et al., 2007).

DcETR1 and *DcERS1* rapidly decreased with the full flower opening and increased at onset of flower senescence. mRNA levels of *DcETR1* and *DcERS1* were up-regulated in the floral organs treated with ethylene while *DcERS2* transcript levels were unaffected by ethylene. Differences between ethylene receptors isoforms have also been observed in other species including *Arabidopsis* (Hua and Meyerowitz, 1998, Binder et al., 2004b and Chen et al., 2007), carnations (Shibuya et al., 2002), roses (Ma et al., 2006), tomato (Tieman et al., 2000 and Kevany et al., 2007), *Delphinium* (Tanase and Ichimura, 2006), and apple (Tatsuki et al., 2009). One potential interpretation is that these increases in gene-specific expression of *DcETR1* and *DcERS1* by ethylene may contribute to an increased sensitivity to ethylene, probably due to the decreased levels of the receptor proteins. However, we cannot rule out that receptor protein levels actually decrease, leading to increased ethylene sensitivity as reported in *Arabidopsis* and tomato (Chen et al., 2007 and Kevany et al., 2007). More refined analyses of receptor protein levels will be required to resolve this.

Overall, our results suggest that ethylene receptors are continually synthesized during later stages of floral development. While ethylene responsiveness is temporarily blocked by 1-MCP treatment, we predict that new receptors synthesized days after 1-MCP treatment can bind ethylene leading to the observed recovery of ethylene sensitivity in the flowers. In addition, we predict that the degradation of ethylene receptors that is stimulated by ethylene binding is prevented by successive treatments with 1-MCP prior to recovery of ethylene sensitivity. This study improves our understanding about the uses and effectiveness of 1-MCP.

5. CONCLUSIONS

Despite the initial block in ethylene perception with 1-MCP, carnation flowers regain ethylene sensitivity and resume endogenous ethylene biosynthesis over time. We were able to demonstrate that recovery of ethylene responsiveness is completely prevented by successive treatments of 1-MCP. In this study, the mRNA levels of *DcETR1* and*DcERS1* decreased during flower opening and increased at onset of flower senescence, suggesting that the receptors are generated during flower development and degraded in senescing tissues. We hypothesize that ethylene receptors are continually synthesized during flower development. Receptors that are newly generated after 1-MCP treatment can bind ethylene which would result in renewed sensitivity to ethylene. Since we observed that successive treatments with 1-MCP completely inhibited ethylene responses, we hypothesize that blocking ethylene binding to the both receptors that is newly synthesized or dissociated from ethylene results in continued inhibition of ethylene responses. By measuring expression in different tissue types, we were able to provide support for this hypothesis. Additional studies on receptor protein levels during similar treatments will be needed to further test this hypothesis.

ACKNOWLEDGEMENTS

This work was funded partially by AgroFresh and the Department of Horticulture, University of Wisconsin-Madison. We would like to thank Dr. Oh-Keun Kwon for technical comments on carnation growing and AgroFresh for supplying 1-MCP.

REFERENCES

1. Abeles, F.B., Morgan, P.W., Saltveit, J.M.E., 1992. Ethylene in Plant Biology. Academic Press, New York.

2. Barry, C.S., Giovannoni, J.J., 2007. Ethylene and fruit ripening. J. Plant Growth Regul. 26, 143–159.

3. Binder, B.M., Bleecker, A.B., 2003. A model for ethylene receptor function and 1- methylcyclopropene action. Acta Hortic. 628, 177–187.

4. Binder, B.M., Mortimore, L.A., Stepanova, A.N., Ecker, J.R., Bleecker, A.B., 2004a. Short-term growth responses to ethylene in Arabidopsis seedlings are EIN3/EIL1 independent. Plant Physiol. 136, 2921–2927.

5. Binder, B.M., O'Malley, R.C., Wang, W., Moore, J.M., Parks, B.M., Spalding, E.P., Bleecker, A.B., 2004b. Arabidopsis seedling growth response and recovery to ethylene. A kinetic analysis. Plant Physiol. 136, 2913–2920.

6. Blankenship, S.M., Dole, J.M., 2003. 1-methylcyclopropene: a review. Postharvest Biol. Technol. 28, 1–25.

7. Borochov, A., Woodson, W.R., 1989. Physiology and Biochemistry of Flower Petal Senescence, Horticultural Reviews. John Wiley & Sons, Inc., pp. 15–43.

8. Cameron, A., Reid, M.S., 2001. 1-MCP blocks ethylene-induced petal abscission of Pelargonium peltatum but the effect is transient. Postharvest Biol. Technol. 22, 169–177.

9. Chang, C., Kwok, S.F., Bleecker, A.B., Meyerowitz, E.M., 1993. Arabidopsis ethyleneresponse gene ETR1: similarity of productto two-component regulators. Science 262, 539–544.

10. Charng,Y.Y., Sun,C.W.,Yan, S.L.,Chou, S.J.,Chen,Y.R.,Yang, S.F., 1997. cDNAsequence of a putative ethylene receptor from carnation petals. Plant Physiol. 115, 863.

11. Chen, Y.F., Shakeel, S.N., Bowers, J., Zhao, X.C., Etheridge, N., Schaller, G.E., 2007. Ligand-induced degradation of the ethylene receptor ETR2 through a proteasome-dependent pathway in Arabidopsis. J. Biol. Chem. 282, 24752–24758.

12. Drory, A., Mayak, S., Woodson, W.R., 1993. Expression of ethylene biosynthetic pathway mRNAs is spatially regulated within carnation flower petals. J. Plant Physiol. 141, 663–667.

13. Ekman, J.H., Clayton, M., Biasi, W.V., Mitcham, E.J., 2004. Interactions between 1- MCP concentration, treatment interval and storage time for 'Bartlett' pears. Postharvest Biol. Technol. 31, 127–136.

14. Halevy, A.H., Mayak, S., 1981. Senescence and postharvest physiology of cut flowers. Part 2. Hortic. Rev. 3, 59–143.

15. Hall, A.E., Findell, J.L., Schaller, G.E., Sisler, E.C., Bleecker, A.B., 2000. Ethylene perception by the ERS1 protein in Arabidopsis. Plant Physiol. 123, 1449–1457.

16. Harada, T., Murakoshi, Y., Torii, Y., Tanase, K., Onozaki, T., Morita, S., Masumura, T., Satoh, S., 2011. Analysis of genomic DNA of DcACS1, a 1-aminocyclopropane-1- carboxylate synthase gene, expressed in senescing petals of carnation (Dianthus caryophyllus) and its orthologous genes in D. superbus var. longicalycinus. Plant Cell Rep. 30, 519–527.

17. Henskens, J.A., Rouwendal, G.J., ten Have, A., Woltering, E.J., 1994. Molecular cloning of two different ACC synthase PCR fragments in carnation flowers and organspecific expression of the corresponding genes. Plant Mol. Biol. 26, 453–458.

18. Hua, J., Meyerowitz, E.M., 1998. Ethylene responses are negatively regulated by a receptor gene family in Arabidopsis thaliana. Cell 94, 261–271.

19. Hua, J., Sakai, H., Nourizadeh, S., Chen, Q.G., Bleecker, A.B., Ecker, J.R., Meyerowitz, E.M., 1998. EIN4 and ERS2 are members of the putative ethylene receptor gene family in Arabidopsis. Plant Cell 10, 1321–1332.

20. Jiang, Y.M., Joyce, D.C., Macnish, A.J., 1999. Responses of banana fruit to treatment with 1-methylcyclopropene. Plant Growth Regul. 28, 77–82.

21. Jones, M.L., Woodson, W.R., 1999. Differential expression of three members of the 1-aminocyclopropane-1-carboxylate synthase gene family in carnation. Plant Physiol. 119, 755–764.

22. Jones, M.L., 2003. Ethylene biosynthetic genes are differentially regulated by ethylene and ACC in carnation styles. Plant Growth Regul. 40, 129–138.

23. Kende, H., 1993. Ethylene biosynthesis. Annu. Rev. Plant Phys. 44, 283–307.

24. Kevany, B.M., Tieman, D.M., Taylor, M.G., Dal Cin, V., Klee, H.J., 2007. Ethylene receptor degradation controls the timing of ripening in tomato fruit. Plant J. 51, 458–467.

25. Kieber, J.J., Rothenberg, M., Roman, G., Feldmann, K.A., Ecker, J.R., 1993. Ctr1, a negative regulator of the ethylene response pathway in arabidopsis, encodes a member of the Raf family of protein-kinases. Cell 72, 427–441.

26. Ma, N., Tan, H.,Xue, J.H., Li, Y.Q., Gao, J.P., 2006. Transcriptional regulation of ethylene receptor and CTR genes involved in ethylene-induced flower opening in cut rose (Rosa hybrida) cv. Samantha. J. Exp. Bot. 57, 2763–2773.

27. Macnish,A.J., Joyce, D.C., Iriving, D.E.,Wearing,A.H., 2004.Asimple sustained release device for the ethylene binding inhibitor 1-methylcyclopropene. Postharvest Biol. Technol. 32, 321–338.

28. Mor, Y., Halevy, A.H., Spiegelstein, H., Mayak, S., 1985. The site of 1- aminocyclopropane-1-carboxylic acid synthesis in senescing carnation petals. Physiol. Plant. 65, 196–202.

29. Nagata, M., Tanikawa, N., Onozaki, T., Mori, H., 2000. Ethylene receptor gene (ETR) homolog from carnation. J. Jpn Soc. Hortic. Sci. 1 (Suppl. 69), 407.

30. O'Malley, R.C., Rodriguez, F.I., Esch, J.J., Binder, B.M., O'Donnell, P., Klee, H.J., Bleecker, A.B., 2005. Ethylene-binding activity, gene expression levels, and receptor system output for ethylene receptor family members from Arabidopsis and tomato. Plant J. 41, 651–659.

31. Overbeek, J.H.M., Woltering, E.J., 1990. Synergistic effect of 1-aminocyclopropane- 1-carboxylic acid and ethylene during

senescence of isolated carnation petals. Physiol. Plant. 79, 368–376.

32. Park, K.Y., Drory, A., Woodson, W.R., 1992. Molecular cloning of an 1- aminocyclopropane-1-carboxylate synthase from senescing carnation flower petals. Plant Mol. Biol. 18, 377–386.

33. Reid, M.S., Wu, M.J., 1992. Ethylene and flower senescence. Plant Growth Regul. 11, 37–43.

34. Reid, M.S., C, elikel, F.G., 2008. Use of 1-methylcyclopropene in ornamentals: carnations as a model system for understanding mode of action. HortScience 43, 95–98.

35. Riov, J., Yang, S.F., 1982. Autoinhibition of ethylene production in citrus peel disks – suppression of 1-aminocyclopropane-1-carboxylic acid synthesis. Plant Physiol. 69, 687–690.

36. Sakai, H., Hua, J., Chen, Q.G., Chang, C., Medrano, L.J., Bleecker, A.B., Meyerowitz, E.M., 1998. ETR2 is an ETR1-like gene involved in ethylene signaling in Arabidopsis. Proc. Natl. Acad. Sci. U.S.A. 95, 5812–5817.

37. Schaller, G.E.,Bleecker,A.B., 1995. Ethylene-binding sites generatedinyeast expressing the Arabidopsis ETR1 gene. Science 270, 1809–1811.

38. Schaller, G.E., 2012. Ethylene and the regulation of plant development. BMC Biol. 10.

39. Schotsmans, W.C., Prange, R.K., Binder, B.M., 2009. 1-Methylcyclopropene: mode of action and relevance in postharvest horticulture research. Hortic. Rev. 35, 263–313.

40. Serek, M., Sisler, E.C., Reid, M.S., 1994. Novel gaseous ethylene binding inhibitor prevents ethylene effects in potted flowering plants. J. Am. Soc. Hortic. Sci. 119, 1230–1233.

41. Serek, M., Sisler, E.C., Reid, M.S., 1995. Effects of 1-MCP on the vase life and ethylene response of cut flowers. Plant Growth Regul. 16, 93–97.

42. Shibuya, K., Satoh, S., Yoshioka, T., 1998. A cDNA endoding a putative ethylene receptor related to petal senescence in carnation (Dianthus caryophyllus L.) flowers. Plant Physiol. 116, 867.

43. Shibuya, K., Nagata, M., Tanikawa, N., Yoshioka, T., Hashiba, T., Satoh, S., 2002. Comparison of mRNA levels of three ethylene receptors in senescing flowers of carnation (Dianthus caryophyllus L.). J. Exp. Bot. 53, 399–406.

44. Sisler, E.C., Reid, M.S., Yang, S.F., 1986. Effect of antagonists of ethylene action on binding of ethylene in cut carnations. Plant Growth Regul. 4, 213–218.

45. Sisler, E.C., Blankenship, S.M., Guest, M., 1990. Competition of cyclooctenes and cyclooctadienes for ethylene binding and activity in plants. Plant Growth Regul. 9, 157–164.

46. Sisler, E.C., Dupille, E., Serek, M., 1996. Effect of 1-methylcyclopropene and methylenecyclopropane on ethylene binding and ethylene action on cut carnations. Plant Growth Regul. 18, 79–86.

47. Sisler, E.C., Serek, M., 2003. Compounds interacting with the ethylene receptor in plants. Plant Biol. 5, 473–480.

48. Tanase, K., Ichimura, K., 2006. Expression of ethylene receptors Dl-ERS1-3 and DlERS2. and ethylene response during flower senescence in Delphinium. J. Plant Physiol. 163, 1159–1166.

49. Tanase, K., Onozaki, T., Satoh, S., Shibata, M., Ichimura, K., 2008. Differential expression levels of ethylene biosynthetic pathway genes during senescence of long-lived carnation cultivars. Postharvest Biol. Technol. 47, 210–217.

50. Tatsuki, M., Hayama, H., Nakamura, Y., 2009. Apple ethylene receptor protein concentrations are affected by ethylene, and differ in cultivars that have different storage life. Planta 230, 407–417.

51. ten Have, A., Woltering, E.J., 1997. Ethylene biosynthetic genes are differentially expressed during carnation (Dianthus caryophyllus L.) flower senescence. Plant Mol. Biol. 34, 89–97.

52. Tieman, D.V., Taylor, M.G., Ciardi, J.A., Klee, H.J., 2000. The tomato ethylene receptors NR and LeETR4 are negative regulators of ethylene response and exhibit functional compensation within a multigene family. Proc. Natl. Acad. Sci. U.S.A. 97, 5663–5668.

53. Verlinden, S., Boatright, J., Woodson, W.R., 2002. Changes in ethylene responsiveness of senescence-related genes during carnation flower development. Physiol. Plant 116, 503–511.

54. Wang, W.Y., Hall, A.E., O'Malley, R., Bleecker, A.B., 2003. Canonical histidine kinase activity of the transmitter domain of the ETR1 ethylene receptor from Arabidopsis is not required for signal transmission. Proc. Natl. Acad. Sci. U.S.A. 100, 352–357.

55. Watkins, C.B., Nock, J.F., Whitaker, B.D., 2000. Responses of early, mid and late season apple cultivars to postharvest application of 1-methylcyclopropene (1- MCP) under air and controlled atmosphere storage conditions. Postharvest Biol. Technol. 19, 17–32.

56. Woodson, W.R., Lawton, K.A., 1988. Ethylene-induced gene-expression in carnation petals – relationship to autocatalytic ethylene production and senescence. Plant Physiol. 87, 498–503.

57. Woodson, W.R., Park, K.Y., Drory, A., Larsen, P.B., Wang, H., 1992. Expression of ethylene biosynthetic pathway transcripts in senescing carnation flowers. Plant Physiol. 99, 526–532.

58. Yang, S.F., Hoffman, N.E., 1984. Ethylene biosynthesis and its regulation in higher plants. Annu. Rev. Plant Phys. 35, 155–189.

INDEX